1,001 Basic Math & Pre-Algebra Practice Problems

FOR DUMMIES®

A Wiley Brand

by Mark Zegarelli

FOR DUMMIES®

A Wiley Brand

1,001 Basic Math & Pre-Algebra Practice Problems For Dummies®

Published by
John Wiley & Sons, Inc.
111 River St.
Hoboken, NJ 07030-5774
www.wiley.com

Library of Congress Control Number: 2013932103

ISBN 978-1-118-44656-0 (pbk); ISBN 978-1-118-44645-4 (ebk); ISBN 978-1-118-44653-9 (ebk); ISBN 978-1-118-44654-6 (ebk)

Manufactured in the United States of America

10 9 8 7 6 5 4 3 2 1

About the Author

Mark Zegarelli is the author of *Basic Math & Pre-Algebra For Dummies, Calculus II For Dummies,* and five other books on math, logic, and test preparation. He holds degrees in both English and math from Rutgers University and is a math tutor and teacher.

Mark lives in San Francisco, California, and Long Branch, New Jersey.

Dedication

This is for Suleiman.

Author's Acknowledgments

This is my eighth *For Dummies* book and, as always, the experience of writing it has been productive and fun. Thanks so much to editors Tim Gallan, Christy Pingleton, Lindsay Lefevere, Shira Fass, and Suzanne Langebartels for setting me right as needed.

And thanks to the folks at Borderlands Café on Valencia Street in San Francisco for the friendly, peaceful, and caffeine-accessible environment that virtually any writer (this one, for example) would find conducive to putting actual words on paper.

Publisher's Acknowledgments

We're proud of this book; please send us your comments at http://dummies.custhelp.com. For other comments, please contact our Customer Care Department within the U.S. at 877-762-2974, outside the U.S. at 317-572-3993, or fax 317-572-4002.

Some of the people who helped bring this book to market include the following:

Acquisitions, Editorial, and Vertical Websites

Senior Project Editor: Tim Gallan

Executive Editor: Lindsay Sandman Lefevere

Copy Editors: Suzanne Langebartels, Christine Pingleton

Assistant Editor: David Lutton

Editorial Program Coordinator: Joe Niesen

Technical Editor: Shira Fass

Editorial Manager: Michelle Hacker

Editorial Assistants: Rachelle S. Amick, Alexa Koschier

Cover Photos: © Sandra van der Steen/iStockphoto.com

Composition Services

Senior Project Coordinator: Kristie Rees

Layout and Graphics: Carrie A. Cesavice, Shawn Frazier, Erin Zeltner

Proofreaders: Jacqui Brownstein, Dwight Ramsey

Indexer: Potomac Indexing, LLC

Publishing and Editorial for Consumer Dummies

   Kathleen Nebenhaus, Vice President and Executive Publisher

   David Palmer, Associate Publisher

   Kristin Ferguson-Wagstaffe, Product Development Director

Publishing for Technology Dummies

   Andy Cummings, Vice President and Publisher

Composition Services

   Debbie Stailey, Director of Composition Services

Contents at a Glance

Table of Contents

Introduction

• •

Are you kidding . . . 1,001 math problems, really?

That's right, a thousand questions plus one to grow on, here in your hot little hands. I've arranged them in order, starting with beginning arithmetic and ending with basic algebra. Topics include everything from the Big Four operations (adding, subtracting, multiplying, and dividing), through negative numbers and fractions, on to geometry and probability, and finally algebra — plus lots more!

Every chapter provides tips for solving the problems in that chapter. And, of course, the back of the book includes detailed explanations of the answers to every question.

It's all here, so get to work!

What You'll Find

This book includes 1,001 basic math and pre-algebra problems, divided into 22 chapters. Each chapter contains problems focusing on a single math topic, such as negative numbers, fractions, or geometry.

Within each chapter, topics are broken into subtopics so that you can work on a specific type of math skill until you feel confident with it. Generally speaking, each section starts with easy problems, moves on to medium ones, and then finishes with hard problems.

You can jump right in anywhere you like and solve these problems in any order. You can also take on one chapter or section at a time, working from easy to medium to hard problems. Or, if you like, you can begin with Question #1 and move right through to Question #1,001.

Additionally, each chapter begins with a list of tips for answering the questions in that chapter.

Every question in Part I is answered in Part II, with a full explanation that walks you through how to understand, set up, and solve the problem.

How This Workbook Is Organized

This workbook includes 1,001 questions in Part I and answers to all of these questions in Part II.

Part 1: Questions

Here are the topics covered by the 1,001 questions in this book:

✔ **Basic arithmetic:** In Chapters 1 through 5, you find dozens of basic arithmetic problems. Chapter 1 begins with rounding numbers and then moves on to basic calculating with addition, subtraction, multiplication, and division. Then, in Chapter 2, you tackle negative numbers, and in Chapter 3, you move on to working with powers and square roots. Chapter 4 gives you plenty of practice in solving arithmetic problems using the order of operations. You may remember this using the mnemonic PEMDAS — **P**arentheses, **E**xponents, **M**ultiplication and **D**ivision, **A**ddition and **S**ubtraction.

Finally, in Chapter 5, you put all of this information together to answer arithmetic word problems, from easy to challenging.

✔ **Divisibility, factors, and multiples:** Chapters 6, 7, and 8 cover a set of topics related to divisibility. In Chapter 6, you discover a variety of divisibility tricks, which allow you to find out whether a number is divisible by another without actually doing the division. You also work with division with remainders and understand the distinction between prime and composite numbers.

Chapter 7 focuses on factors and multiples. You discover how to generate all the factors and prime factors of a number and calculate the greatest common factor (GCF) for a set of two or more numbers. Additionally, you generate a partial list of the multiples, and calculate the least common multiple (LCM) of two or more numbers.

Chapter 8 wraps up the section with word problems that sharpen and extend your skills working with factors, multiples, remainders, and prime numbers.

✔ **Fractions, decimals, percents, and ratios:** Chapters 9 through 13 focus on four distinct ways to represent parts of a whole — fractions, decimals, percents, and ratios. In Chapter 9, you work with fractions, including increasing the terms of fractions and reducing them to lowest terms. You change improper fractions to mixed numbers, and vice versa. You add, subtract, multiply, and divide fractions, including mixed numbers. You also simplify complex fractions.

In Chapter 10, you convert fractions to decimals, and vice versa. You add, subtract, multiply, and divide decimals. You also find out how to work with repeating decimals. Chapter 11 focuses on percents. You convert fractions and multiples to percents, and vice versa. You discover a few tricks for calculating simple percents. You also work on more difficult percent problems by creating word equations, which can then be translated into equations and solved.

Chapter 12 presents a variety of problems, including word problems, that use ratios and proportions. And in Chapter 13, you tackle even more word problems where you apply your skills working with fractions, decimals, and percents.

✔ **Scientific notation, weights and measures, geometry, graphs, statistics and probability, and sets:** In Chapters 14 through 19, you take a great stride forward working with a wide variety of intermediate basic math skills. In Chapter 14, the topic is scientific notation, which is used to represent very large and very small numbers. Chapter 15 introduces you to weights and measures, focusing on the English and metric systems, and conversions between the two. Chapter 16 gives you a huge number of geometry problems of every description, including both plane and solid geometry. In Chapter 17, you work with a variety of graphs, including bar graphs, pie charts, line graphs, pictographs, and the *xy*-graph that is used so much in algebra and later math.

Chapter 18 gives you an introduction to basic statistics, including the mean, median, and mode. It also provides problems in probability and gives you an introduction to counting both independent and dependent events. Chapter 19 gives you some problems in basic set theory, including finding the union, intersection, relative complement, and complement. Additionally, you use Venn diagrams to solve word problems.

✔ **Algebraic expressions and equations:** To finish up, Chapters 20, 21, and 22 give you a taste of the work you'll be doing in your first algebra class. Chapter 20 shows you the basics of working with algebraic expressions, including evaluating, simplifying, and factoring. In Chapter 21, you solve basic algebraic equations. And in Chapter 22, you put these skills to use, solving a set of word problems with basic algebra.

Part II: Answers

In this part, you find answers to all 1,001 questions that appear in Part I. Each answer contains a complete step-by-step explanation of how to solve the problem from beginning to end.

Beyond the Book

This book gives you plenty of math to work on and prepares you for algebra. But maybe you want to track your progress as you tackle the problems, or maybe you're having trouble with certain types of problems and wish they were all presented in one place where you could methodically make your way through them. You're in luck. Your book purchase comes with a free one-year subscription to all 1,001 practice problems online. You get on-the-go access any way you want it — from your computer, smartphone, or tablet. Track your progress and view personalized reports that show where you need to study the most. And then do it. Study what, where, when, and how you want.

What you'll find online

The online practice that comes free with this book offers you the same 1,001 questions and answers that are available here, presented in a multiple-choice format. The beauty of the online problems is that you can customize your online practice to focus on the topic areas that give you the most trouble. So if you need help converting fractions to decimals or have trouble grasping pre-algebra, then select these problem types online and start practicing. Or, if you're short on time but want to get a mixed bag of a limited number of problems, you can specify the quantity of problems you want to practice. Whether you practice a few hundred problems in one sitting or a couple dozen, and whether you focus on a few types of problems or practice every type, the online program keeps track of the questions you get right and wrong so that you can monitor your progress and spend time studying exactly what you need.

You can access this online tool using a PIN code, as described in the next section. Keep in mind that you can create only one login with your PIN. Once the PIN is used, it's no longer valid and is nontransferable. So you can't share your PIN with other users after you've established your login credentials.

How to register

Purchasing this book entitles you to one year of free access to the online, multiple-choice version of all 1,001 of this book's practice problems. All you have to do is register. Just follow these simple steps:

1. **Find your PIN code.**

   • **Print book users:** If you purchased a hard copy of this book, turn to the back of this book to find your PIN.

   • **E-book users:** If you purchased this book as an e-book, you can get your PIN by registering your e-book at dummies.com/go/getaccess. Go to this website, find your book and click it, and then answer the security question to verify your purchase. Then you'll receive an e-mail with your PIN.

2. **Go to onlinepractice.dummies.com.**

3. **Enter your PIN.**

4. **Follow the instructions to create an account and establish your own login information.**

That's all there is to it! You can come back to the online program again and again — simply log in with the username and password you choose during your initial login. No need to use the PIN a second time.

If you have trouble with the PIN or can't find it, please contact Wiley Product Technical Support at 800-762-2974 or http://support.wiley.com.

Your registration is good for one year from the day you activate your PIN. After that time frame has passed, you can renew your registration for a fee. The website gives you all the important details about how to do so.

Where to Go for Additional Help

Every chapter in this book opens with tips for solving the problems in that chapter. And, of course, if you get stuck on any question, you can flip to the answer section and try to work through the solution provided. However, if you feel that you need a bit more basic math information than this book provides, I highly recommend my earlier book *Basic Math & Pre-Algebra For Dummies*. This book gives you a ton of useful information for solving every type of problem included here.

Additionally, you can also check out my *Basic Math & Pre-Algebra Workbook For Dummies*. It contains a nice mix of short explanations for how to do various types of problems, followed by practice. And, for a quick take on the most important basic math concepts, have a look at *Basic Math & Pre-Algebra Essentials For Dummies*. Yep, I wrote that one, too — how's that for shameless plugs?

Part I
The Questions

1001 Questions

In this part . . .

One thousand and one math problems. That's one problem for every night in the *Arabian Nights* stories. That's almost ten problems for every floor in the Empire State Building. In short, that's a *lot* of problems — plenty of practice to help you attain the math skills you need to do well in your current math class. Here's an overview of the types of questions provided:

✔ Basic arithmetic, including absolute value, negative numbers, powers, and square roots (Chapters 1–5)

✔ Divisibility, factors, and multiples (Chapters 6–8)

✔ Fractions, decimals, percents, and ratios (Chapters 9–13)

✔ Scientific notation, measures, geometry, graphs, statistics, probability, and sets (Chapters 14–19)

✔ Algebraic expressions and equations (Chapters 20–22)

Chapter 1

The Big Four Operations

The Big Four operations (adding, subtracting, multiplying, and dividing) are the basis for all of arithmetic. In this chapter, you get plenty of practice working with these important operations.

The Problems You'll Work On

Here are the types of problems you find in this chapter:

- ✔ Rounding numbers to the nearest ten, hundred, thousand, or million
- ✔ Adding columns of figures, including addition with carrying
- ✔ Subtracting one number from another, including subtraction with borrowing
- ✔ Multiplying one number by another
- ✔ Division, including division with a remainder

What to Watch Out For

Here's a quick tip for rounding numbers to help you in this chapter: When rounding a number, check the number to the right of the place you're rounding to. If that number is from 0 to 4, round down by changing that number to 0. If that number is from 5 to 9, round up by changing that number to 0 and adding 1 to the number to its left.

For example, to round 7,654 to the nearest hundred, check the number to the right of the hundreds place. That number is 5, so change it to 0 and add 1 to the 6 that's to the left of it. Thus, 7,654 becomes 7,700.

Rounding

1–6

1. Round the number 136 to the nearest ten.

2. Round the number 224 to the nearest ten.

3. Round the number 2,492 to the nearest hundred.

4. Round the number 909,090 to the nearest hundred.

5. Round the number 9,099 to the nearest thousand.

6. Round the number 234,567,890 to the nearest million.

Adding, Subtracting, Multiplying, and Dividing

7–30

7. Add $47 + 21 = ?$

8. Add $136 + 53 + 77 = ?$

9. Add $735 + 246 + 1,329 = ?$

10. Add $904 + 1,024 + 6,532 + 883 = ?$

11. Add $56,702 + 821 + 5,332 + 89 + 343,111 = ?$

12. Add $1,609,432 + 657,936 + 82,844 + 2,579 + 459 = ?$

13. Subtract $89 - 54 = ?$

14. Subtract $373 - 52 = ?$

15. Subtract $539 - 367 = ?$

16. Subtract $2,468 - 291 = ?$

17. Subtract $34,825 - 26,492 = ?$

18. Subtract $71,002 - 56,234 = ?$

19. Multiply $458 \times 4 = ?$

20. Multiply $74 \times 35 = ?$

21. Multiply $129 \times 86 = ?$

22. Multiply $382 \times 67 = ?$

23. Multiply $9,876 \times 34 = ?$

24. Multiply $23,834 \times 1,597 = ?$

25. Divide $861 \div 3 = ?$

26. Divide $1,876 \div 7 = ?$

27. Divide $6,184 \div 15 = ?$

28. Divide $25,246 \div 22 = ?$

29. Divide $60,000 \div 53 = ?$

30. Divide $262,145 \div 256 = ?$

Chapter 2

Less than Zero: Working with Negative Numbers

Negative numbers can be a cause of negativity for some students. The rules for working with negative numbers can be a little tricky. In this chapter, you practice applying the Big Four operations to negative numbers. You also strengthen your skills evaluating absolute value.

The Problems You'll Work On

This chapter shows you how to work with the following types of problems:

- Subtracting a smaller number minus a larger number
- Adding and subtracting with negative numbers
- Multiplying and dividing with negative numbers
- Evaluating absolute value

What to Watch Out For

Here are a few things to keep an eye out for when you're working with negative numbers:

- To subtract a smaller number minus a larger number, reverse and negate: *Reverse* by subtracting the larger number minus the smaller one, and then *negate* by attaching a minus sign (–) in front of the result. For example, $4 - 7 = -3$.
- To subtract a negative number minus a positive number, add and negate: *Add* the two numbers as if they were positive, then *negate* by attaching a minus sign in front of the result. For example, $-5 - 4 = -9$.
- To add a positive number and a negative number (in either order), subtract the larger number minus the smaller number; then attach the same sign to the result as the number that is farther from 0. For example, $-3 + 5 = 2$ and $4 + (-6) = -2$.

Adding and Subtracting Negative Numbers

31–41

31. Evaluate each of the following.

i. $3 - 6 =$

ii. $7 - 12 =$

iii. $14 - 15 =$

iv. $2 - 16 =$

v. $20 - 31 =$

32. Evaluate each of the following.

i. $-7 - 4 =$

ii. $-1 - 9 =$

iii. $-9 - 6 =$

iv. $-11 - 6 =$

v. $-1 - 13 =$

33. Evaluate each of the following.

   i. $-5+8=$

   ii. $-8+5=$

   iii. $-14+1=$

   iv. $-1+14=$

   v. $-20+6=$

34. Evaluate each of the following.

   i. $-2+(-8)=$

   ii. $6+(-3)=$

   iii. $-9+(-3)=$

   iv. $15+(-5)=$

   v. $-19+(-1)=$

35. Evaluate each of the following.

   i. $4-(-2)=$

   ii. $-9-(-1)=$

   iii. $-10-(-3)=$

   iv. $8-(-11)=$

   v. $-3-(-16)=$

36. $-29+(-35)=$

37. $46-(-89)=$

38. $81+(-137)=$

39. $-212-942=$

40. $1{,}024-2{,}543=$

41. $-10{,}654-(-289)=$

Multiplying and Dividing Negative Numbers

42–53

42. Evaluate each of the following.

   i. $-6\times9=$

   ii. $-8\times(-7)=$

iii. $-9 \times (-7) =$

47. $2 \times (-4) \times (-10) \times (-5) =$

iv. $7 \times (-8) =$

48. $-1 \times (-2) \times 3 \times (-4) \times (-5) \times (-1) =$

v. $-9 \times (-6) =$

49. Evaluate each of the following.

43. $-15 \times 9 =$

    i. $35 \div (-5) =$

44. $-32 \times (-11) =$

    ii. $-28 \div (-4) =$

45. $91 \times (-18) =$

    iii. $32 \div (-4) =$

46. $-7 \times (-6) \times 5 =$

    iv. $-48 \div -6 =$

    v. $-36 \div 6 =$

50. $176 \div (-8) =$

51. $-403 \div 13 =$

52. $-275 \div (-11) =$

53. $-1,054 \div (-17) =$

Working with Absolute Value

54–57

54. Evaluate each of the following.

   i. $|4-4| =$

  ii. $|6-2| =$

  iii. $|7-9| =$

  iv. $|9-1| =$

  v. $|1-8| =$

55. $|38-99| =$

56. $|206-88| =$

57. $|543-629| =$

Chapter 3

You've Got the Power: Powers and Roots

$\mathcal{P}$owers provide a shorthand notation for multiplication using a base number and an exponent. Roots — also called radicals — reverse the process of powers. In this chapter, you practice taking powers and roots of positive integers as well as fractions and negative integers.

The Problems You'll Work On

This chapter deals with the following types of problems:

- ✔ Using powers to multiply a number by itself
- ✔ Applying exponents to negative numbers and fractions
- ✔ Understanding square roots
- ✔ Knowing how to evaluate negative exponents and fractional exponents

What to Watch Out For

Following are some tips for working with powers and roots:

- ✔ When you find the power of a number, multiply the base by itself as many times as indicated by the exponent. For example, $4^3 = 4 \times 4 \times 4 = 64$.

- ✔ When the base is a negative number, use the standard rules of multiplication for negative numbers (see Chapter 2). For example, $-7^2 = -7 \times (-7) = 49$.

- ✔ When the base is a fraction, use the standard rules of multiplication for fractions (see Chapter 9). For example, $\left(\frac{2}{5}\right)^3 = \frac{2}{5} \times \frac{2}{5} \times \frac{2}{5} = \frac{8}{125}$.

- ✔ To find the square root of a square number, find the number that, when multiplied by itself, results in the number you started with. For example, $\sqrt{36} = 6$, because $6 \times 6 = 36$.

- ✔ To simplify the square root of a number that's not a square number, if possible, factor out a square number and then evaluate it. For example, $\sqrt{12} = \sqrt{4}\sqrt{3} = 2\sqrt{3}$.

- ✔ Evaluate an exponent of $\frac{1}{2}$ as the square root of the base. For example, $25^{\frac{1}{2}} = \sqrt{25} = 5$.

- ✔ Evaluate an exponent of –1 as the reciprocal of the base. For example, $7^{-1} = \frac{1}{7}$.

- ✔ To evaluate an exponent of a negative number, make the exponent positive and evaluate its reciprocal. For example, $3^{-2} = \frac{1}{3^2} = \frac{1}{9}$.

Multiplying a Number by Itself

58–72

58. Evaluate each of the following.

    i. 6^2

    ii. 12^2

    iii. 2^6

    iv. 3^4

    v. 71^0

59. $26^2 =$

60. $12^3 =$

61. $10^6 =$

62. $20^5 =$

63. $100^4 =$

64. $101^3 =$

65. Evaluate each of the following.

   i. $(-5)^2$

   ii. $(-4)^3$

   iii. $(-10)^5$

   iv. $(-1)^{12}$

   v. $(-1)^{27}$

66. $(-11)^4 =$

67. $(-15)^3 =$

68. $(-40)^5 =$

69. Evaluate each of the following.

   i. $\left(\dfrac{1}{6}\right)^2$

   ii. $\left(\dfrac{1}{3}\right)^3$

   iii. $\left(\dfrac{7}{11}\right)^2$

   iv. $\left(\dfrac{2}{5}\right)^4$

   v. $\left(\dfrac{1}{10}\right)^5$

70. $\left(\dfrac{9}{22}\right)^2 =$

71. $\left(\frac{7}{30}\right)^3 =$

72. $\left(\frac{2}{3}\right)^7 =$

Finding Square Roots

73–79

73. Simplify each of the following as a whole number by finding the square root.

   i. $\sqrt{9}$

   ii. $\sqrt{36}$

   iii. $\sqrt{64}$

   iv. $\sqrt{144}$

   v. $\sqrt{289}$

74. Simplify each of the following as a whole number by finding the square root and then multiplying.

   i. $2\sqrt{16}$

   ii. $3\sqrt{25}$

   iii. $6\sqrt{100}$

   iv. $9\sqrt{121}$

   v. $20\sqrt{225}$

75. $\sqrt{8} =$

76. $\sqrt{32} =$

77. $\sqrt{54} =$

78. $\sqrt{80} =$

79. $\sqrt{300} =$

Negative and Fractional Exponents

80–90

80. Express each of the following as a square root and then simplify as a positive whole number.

i. $4^{\frac{1}{2}}$

ii. $49^{\frac{1}{2}}$

iii. $81^{\frac{1}{2}}$

iv. $169^{\frac{1}{2}}$

v. $400^{\frac{1}{2}}$

81. $27^{\frac{1}{2}} =$

82. $52^{\frac{1}{2}} =$

83. $72^{\frac{1}{2}} =$

84. $99^{\frac{1}{2}} =$

85. Simplify each of the following as a fraction.

86. $7^{-2} =$

   i. 3^{-1}

87. $2^{-6} =$

   ii. 4^{-1}

88. $5^{-4} =$

   iii. 10^{-1}

89. $13^{-2} =$

   iv. 16^{-1}

90. $10^{-6} =$

   v. 100^{-1}

Chapter 4

Following Orders: Order of Operations

The order of operations (also called the order of precedence) provides a clear way to evaluate complex expressions so you always get the right answer. The mnemonic PEMDAS helps you to remember to evaluate parentheses first; then move on to exponents; then multiplication and division; and finally addition and subtraction.

The Problems You'll Work On

This chapter includes these types of problems:

- Evaluating expressions that contain the Big Four operations (addition, subtraction, multiplication, and division)
- Evaluating expressions that include exponents
- Evaluating expressions that include parentheses, including nested parentheses
- Evaluating expressions that include parenthetical expressions, such as square roots and absolute value
- Evaluating expressions that include fractions with expressions in the numerator and/or denominator

What to Watch Out For

Keep the following tips in mind as you work with the problems in this chapter:

- When an expression has only addition and subtraction, evaluate it from left to right. For example, $8 - 5 + 6 = 3 + 6 = 9$.
- When an expression has only multiplication and division, evaluate it from left to right. For example, $10 \div 2 \times 7 = 5 \times 7 = 35$.
- When an expression has any combination of the Big Four operations, first evaluate all multiplication and division from left to right; then evaluate addition and subtraction from left to right. For example, $8 + 12 \div 4 = 8 + 3 = 11$.
- When an expression includes powers, evaluate them *first*, and *then* evaluate Big Four operations. For example, $4 - 3^2 = 4 - 9 = -5$.

The Big Four Operations

91–102

91. $8 + 9 - 3 =$

92. $-5 - 10 + 3 - 4 =$

93. $4 \times 6 \div 8 =$

94. $28 \div 7 \times 4 \div 2 =$

95. $-35 \div 7 \times (-6) =$

96. $72 \div (-9) \times (-4) \div 2 =$

97. $56 \div 7 + 1 =$

98. $15 - 8 \times 2 =$

99. $12 + 10 \div 2 - 1 =$

100. $18 + 36 \div 9 \times 2 =$

101. $75 \div (-5) \times 3 + 4 =$

102. $-6 \times 7 + (-36) \div 3 =$

Operations with Exponents

103–112

103. $4 \times 10^2 =$

104. $56 \div 2^3 \times 20 =$

105. $1 + 5^2 - 4 =$

106. $3^3 + 2^3 - 10 =$

107. $-2^5 + 3^2 =$

108. $7^2 - 6^0 \times 3 =$

109. $10^5 \div 10^4 - 10 =$

110. $-20 \times 25 + 2^3 \times 5^3 =$

111. $(-8)^2 \div 2^3 \times 40 + (-200) =$

112. $-1^3 \times (-2) + 9^2 \div 3^3 =$

Operations with Parentheses

113–124

113. $7^2 \times (6 - 3) =$

114. $5 \times (3 - 9 \times 2) =$

115. $(-9 \div 3) \div \left((-6)^2 \div 12\right) =$

116. $(5 \times 3 - 1) \times \left(50 \div 5^2\right) =$

117. $(11 - 3)^2 \div \left(6^2 - 4\right) =$

118. $\left[12 \div (-4)\right] \times \left(10 \times 2^2 + 1\right) =$

119. $\left(-1^3 - 5\right)^2 - \left(6 - 4 \div 2\right)^2 =$

120. $\left[(5-2)\times 4\right]+1=$

121. $50-\left[(-6+2)\times 3^2\right]=$

122. $3\times\left[4\times(-3+8)^2\right]=$

123. $-24\div\left[(-2-10)\times\left(7-2^3\right)\right]=$

124. $4+\left\{\left[(5-1)\times 7\right]\div 14\right\}=$

Operations with Square Roots

125–134

125. $\sqrt{64}\div 4=$

126. $\sqrt{100}-\sqrt{36}=$

127. $-1+\sqrt{81}\div 9=$

128. $\sqrt{4\times 9}\div 2+4=$

129. $-8-\sqrt{24+3\div 3}=$

130. $\sqrt{-20\div(-7+2)}+5=$

131. $\sqrt{79-5\times 2^4}+2=$

132. $\sqrt{5^2\times 3^2}+\sqrt{5^2-3^2}=$

133. $\left[\sqrt{(13+5)\times 2}-4^2\right]^2=$

134. $\left(\sqrt{4+\sqrt{4^2\times 2^4}\times 2}-3^2\right)^3=$

Operations with Fractions

135–140

135. $\dfrac{8-2}{16\div 8} =$

136. $\dfrac{4\times\sqrt{25}}{-7-(-2)} =$

137. $\dfrac{2^5-2^4+2^3}{2^4+2^3} =$

138. $\dfrac{3^3+17}{-6+(-8\times 2)} =$

139. $\dfrac{\sqrt{2^5-(-4)}}{\left[12-(1-7)\right]\div 6} =$

140. $\sqrt{\dfrac{\left[22\div\left(7+2^2\right)\right]+\left(7^2-15\right)}{\sqrt{4^3+2^4-\left(-1^3\right)}}} =$

Operations with Absolute Values

141–144

141. $\left|-8+2\times(-5)\right|\div(-3) =$

142. $\left|(7-11)\div 2\right|\times(3-13) =$

143. $\left|4-9\right|\times(17-5)-\left|8\div(-2)\right| =$

144. $\dfrac{\sqrt{\left|44-85\right|+\left|(5-70)\div 13\right|-(-3)}}{\left[7\div(10-3)\right]+(-8)} =$

Chapter 5
Big Four Word Problems

. . . .

Word problems provide an opportunity for you to apply your math skills to real-world situations. In this chapter, all the problems can be solved using the Big Four operations (adding, subtracting, multiplying, and dividing).

The Problems You'll Work On

The problems in this chapter fall into three basic categories, based on their difficulty:

- ✔ Basic word problems where you need to perform a single operation
- ✔ Intermediate word problems where you need to use two different operations
- ✔ Tricky word problems that require several different operations and more difficult calculations

What to Watch Out For

Here are a few tips for getting the right answer to word problems:

- ✔ Read each problem carefully to make sure you understand what it's asking.
- ✔ Use scratch paper to gather and organize information from the problem.
- ✔ Think about which Big Four operation (adding, subtracting, multiplying, or dividing) will be most helpful for solving the problem.
- ✔ Perform calculations carefully to avoid mistakes.
- ✔ Ask yourself whether the answer you got makes sense.
- ✔ Check your work to make sure you're right.

Basic Word Problems

145–154

145. A horror movie triple-feature included *Zombies Are Forever*, which was 80 minutes long, *An American Werewolf in Bermuda*, which ran for 95 minutes, and *Late Night Snack of the Vampire*, which was 115 minutes from start to finish. What was the total length of the three movies?

146. At a height of 2,717 feet, the tallest building in the world is the Burj Khalifa in Dubai. It's 1,263 feet taller than the Empire State Building in New York City. What is the height of the Empire State Building?

147. Janey's six children are making colored eggs for Easter. She bought a total of five dozen eggs for all of the children to use. Assuming each child gets the same number of eggs, how many eggs does each child receive?

148. Arturo worked a 40-hour week at $12 per hour. He then received a raise of $1 per hour and worked a 30-hour week. How much more money did he receive for the first week of work than the second?

149. A restaurant has 5 tables that seat 8 people each, 16 tables with room for 6 people each, and 11 tables with room for 4 people each. What is the total capacity of all the tables at the restaurant?

150. The word *pint* originally comes from the word *pound* because a pint of water weighs 1 pound. If a gallon contains 8 pints, how many pounds does 40 gallons of water weigh?

151. Antonia purchased a sweater normally priced at $86, including tax. When she brought it to the cash register, she found that it was selling for half off. Additionally, she used a $20 gift card to help pay for the purchase. How much money did she have to spend to buy the sweater?

152. A large notebook costs $1.50 more than a small notebook. Karan bought two large notebooks and four small notebooks, while Almonte bought five large notebooks and one small notebook. How much more did Almonte spend than Karan?

153. A company invests $7,000,000 in the development of a product. Once the product is on the market, each sale returns $35 on the investment. If the product sells at a steady rate of 25,000 per month, how long will it take for the company to break even on its initial investment?

154. Jessica wants to buy 40 pens. A pack of 8 pens costs $7, but a pack of 10 pens costs $8. How much does she save by buying packs of 10 pens instead of packs of 8 pens?

Intermediate Word Problems

155–171

155. Jim bought four boxes of cereal on sale. One box weighed 10 ounces and the remaining boxes weighed 16 ounces each. How many ounces of cereal did he buy altogether?

156. Mina took a long walk on the beach each day of her eight-day vacation. On half of the days, she walked 3 miles and on the other half she walked 5 miles. How many miles did she walk altogether?

157. A three-day bike-a-thon requires riders to travel 100 miles on the first day and 20 miles fewer on the second day. If the total trip is 250 miles, how many miles do they travel on the third day?

158. If six T-shirts sell for $42, what is the cost of nine T-shirts at the same rate?

159. Kenny did 25 pushups. His older brother, Sal, did twice as many pushups as Kenny. Then, their oldest sister, Natalie, did 10 more pushups than Sal. How many pushups did the three children do altogether?

160. A candy bar usually sells at two for 90 cents. This week, it is specially packaged at three for $1.05. How much can you save on a single candy bar by buying a package of three rather than two?

161. Simon noticed a pair of square numbers that add up to 130. He then noticed that when you subtract one of these square numbers from the other, the result is 32. What is the smaller of these two square numbers?

162. If Donna took 20 minutes to read 60 pages of a 288-page graphic novel, how long did she take to read the whole novel, assuming that she read it all at the same rate?

163. Kendra sold 50 boxes of cookies in 20 days. Her older sister, Alicia, sold twice as many boxes in half as many days. If the two girls continued at the same sales rates, how many total boxes would both girls have sold if they had both sold cookies for 40 days?

164. A group of 70 third graders has exactly three girls for every four boys. When the teacher asks the children to pair up for an exercise, six boy-girl pairs are formed, and the rest of the children pair up with another child of the same sex. How many more boy-boy pairs are there than girl-girl pairs?

165. Together, a book and a newspaper cost $11.00. The book costs $10.00 more than the newspaper. How many newspapers could you buy for the same price as the book?

166. Yianni just purchased a house priced at $385,000 with a mortgage from the bank. His monthly mortgage payment to cover the principal and interest will be $1,800 per month for 30 years. When he has finished paying off the house, how much over and above the cost of the house will Yianni have paid in interest?

167. The distance from New York to San Diego is approximately 2,700 miles. Because of prevailing winds, when flying east-to-west, the flight usually takes one hour longer than when flying west-to-east. If a plane from San Diego to New York travels at a forward speed of 540 miles per hour, what is the forward speed of a plane traveling from New York to San Diego under the same conditions?

168. Arlo went to an all-night poker game hosted by friends. By 11:00, he was down $65 from where he had started. Between 11:00 and 2:00, he won $120. Then, in the next three hours, he lost another $45. In the final hour of the game, he won $30. How much did Arlo win or lose during the game?

169. Clarissa bought a diamond for $1,000 and then sold it to Andre for $1,100. A month later, Andre needed money, so he sold the diamond back to Clarissa for $900. But a few months later, he had a windfall and bought the diamond back from Clarissa for $1,200. How much profit did Clarissa make as a result of the total transactions?

170. Angela and Basil both work at a cafeteria making sandwiches. At top speed, Angela can make four sandwiches in three minutes and Basil can make three sandwiches in four minutes. Working together, how long will they take to make 200 sandwiches?

171. All 16 children in Ms. Morrow's preschool have either two or three siblings. Altogether, the children have a total of 41 siblings. How many of the children have three siblings?

Advanced Word Problems

172–180

172. What is the sum of all the numbers from 1 to 100?

173. Louise works in retail and has a $1,200-per-day sales quota. On Monday, she exceeded this quota by $450. On Tuesday, she exceeded it by $650. On Wednesday and Thursday, she made her quota exactly. Friday was a slow day, so Louise sold $250 less than her quota. What were her total sales for the five days?

174. A sign posted over a large swimming pool reminds swimmers that 40 lengths of the pool equals 1 mile. Jordy swam 1 length of the pool at a rate of 3 miles per hour. How long did he take to swim 1 length of the pool?

175. In a group of two people, only one pair can shake hands. But in a group of three people, three different pairings of people can shake hands. How many different pairings of people can shake hands in a group of ten people?

176. Marion found that three red bricks and one white brick weighed a total of 23 pounds. Then she replaced one red brick on the scale with two white bricks, and found that the weight went up to 27 pounds. Assuming all red bricks are equal to each other in weight, and that the same is true of all white bricks, what is the weight of one red brick?

177. Angela counted all the coins in her uncle's change jar. She counted 891 pennies, 342 nickels, 176 dimes, and 67 quarters. How much money was in the jar?

178. On a long car trip, Joel drove the first two hours on the highway at 70 miles per hour. He took a 15-minute break and then drove another hour at 60 miles per hour. Next, he drove on a winding mountain road for two hours at 35 miles per hour. He took a 45-minute break for dinner and then finished up his trip with three hours of driving at 75 miles per hour. What was Joel's average speed for the whole trip, including the breaks?

179. A candy bar usually sells at two for 90 cents. This week, it is specially packaged at three for $1.05. Heidi bought so many that she saved $5.40 in comparison with the regular price. How many candy bars did she buy?

180. Suppose you decide to save a dollar on the first day of the month, two additional dollars on the second, four additional dollars the third, and continue doubling the amount you add every day. How many days will it take you to save a combined total of more than $30,000?

Chapter 6

Divided We Stand

● ●

Division is the most interesting and complex of the Big Four operations (addition, subtraction, multiplication, and division). When you divide two numbers, you divide the *dividend* by the *divisor,* and the result is the *quotient.* For example:

Dividend		Divisor		Quotient
24	÷	8	=	3

Integer division — that is, division with whole numbers only — always results in a remainder (which may be 0).

Dividend		Divisor		Quotient	Remainder
26	÷	8	=	3	*r*2

The remainder is a whole number from 0 to one less than the number you're dividing by. For example, when you divide any number by 8, the remainder must be a whole number from 0 to 7, inclusive.

The Problems You'll Work On

This chapter focuses on the following concepts and skills:

- ✔ Understanding integer division — that is, division with a remainder
- ✔ Knowing some quick rules for divisibility
- ✔ Finding the remainder to a division problem without dividing
- ✔ Distinguishing prime numbers from composite numbers

What to Watch Out For

Following are some rules and tips to utilize when working division problems:

- ✔ One integer is *divisible* by another when the result of division is a remainder of 0. For example, $54 \div 3 = 18r0$, so 54 is divisible by 3.
- ✔ A *prime* number is divisible by exactly two numbers: 1 and the number itself. For example, 17 is a prime number because it's divisible only by 1 and 17.
- ✔ A *composite* number is divisible by 3 or more numbers. For example, 25 is a composite number because it's divisible by 1, 5, and 25.
- ✔ The number 1 is the only number that is neither prime nor composite.

Determining Divisibility

181–200

181. Which of the following numbers are divisible by 2?

   i.   32

   ii. 70

   iii. 109

   iv. 8,645

   v. 231,996

182. Which of the following numbers are divisible by 3?

   i.   51

   ii. 77

   iii. 138

   iv. 1,998

   v. 100,111

183. Which of the following numbers are divisible by 4?

i. 57

ii. 552

iii. 904

iv. 12,332

v. 7,435,830

184. Which of the following numbers are divisible by 5?

i. 190

ii. 723

iii. 1,005

iv. 252,525

v. 505,009

185. Which of the following numbers are divisible by 6?

    i. 61

    ii. 88

    iii. 372

    iv. 8,004

    v. 1,001,010

186. Which of the following numbers are divisible by 8?

    i. 881

    ii. 1,914

    iii. 39,888

    iv. 711,124

    v. 43,729,408

187. Which of the following numbers are divisible by 9?

   i.   98

   iii.   324

   iii.   6,009

   iv.   54,321

   v.   993,996

188. Which of the following numbers are divisible by 10?

   i.   340

   ii.   8,245

   iii.   54,002

   iv.   600,010

   v.   1,010,100

189. Which of the following numbers are divisible by 11?

   i. 134

   ii. 209

   iii. 681

   iv. 1,925

   v. 81,928

190. Which of the following numbers are divisible by 12?

   i. 81

   ii. 132

   iii. 616

   iv. 123,456

   v. 12,345,678

191. What's the greatest power of 10 that's a factor of 87,000?

192. What's the greatest power of 10 that's a factor of 9,200,000?

193. What's the greatest power of 10 that's a factor of 30,940,050?

194. The number 78 is divisible by which numbers from 2 to 6, inclusive? (*Note:* More than one answer is possible.)

195. The number 128 is divisible by which numbers from 2 to 6, inclusive? (*Note:* More than one answer is possible.)

196. The number 380 is divisible by which numbers from 2 to 6, inclusive? (*Note:* More than one answer is possible.)

197. The number 6,915 is divisible by which numbers from 2 to 6, inclusive? (*Note:* More than one answer is possible.)

198. You know that the number 56 is divisible by 7 (because $56 \div 7 = 8$.) Using that information, what is the remainder when you divide $59 \div 7$?

199. You know that the number 612 is divisible by 9 (because $6 + 1 + 2 = 9$). Using that information, what is the remainder when you divide $611 \div 9$?

200. You know that the number 9,000 is divisible by 6 (because it's an even number whose digits add up to 9, which is divisible by 3). Using that information, what is the remainder when you divide $8,995 \div 6$?

Working with Prime and Composite Numbers

201–210

201. Which of the following are prime numbers and which are composite numbers?

    i. 39

    ii. 41

    iii. 57

    iv. 73

    v. 91

202. Is 143 a prime number?

203. Is 151 a prime number?

204. Is 161 a prime number?

205. Is 223 a prime number?

206. Is 267 a prime number?

207. Which two different prime numbers is 93 divisible by?

208. Which two different prime numbers is 297 divisible by?

209. Which two different prime numbers is 448 divisible by?

210. Which three prime numbers is 293,425 divisible by?

Chapter 7

Factors and Multiples

• •

Factors and multiples are two key mathematical ideas that are both related to division. When one number is divisible by another number, the first number is a *multiple* of the second, and the second is a *factor* of the first. Calculating the factors and multiples of a number is essential for the more complicated math that follows, such as working with fractions, decimals, and percents (which are covered in Chapters 9 through 11).

The Problems You'll Work On

The following are the types of problems to which you apply your math skills in this chapter:

✔ Deciding whether one number is a factor of another number

✔ Generating all the factors of a number

✔ Finding the prime factors of a number

✔ Discovering the greatest common factor (GCF) of two or more numbers

✔ Listing the first few multiples of a number

✔ Finding the least common multiple (LCM) of two or more numbers

What to Watch Out For

Here are a few tips for handling the problems you find in this chapter:

✔ When one number is divisible by a second number, the second number is a *factor* of the first. For example, 10 is divisible by 5, so 5 is a factor of 10.

✔ When one number is divisible by a second number, the first number is a *multiple* of the second. For example, 10 is divisible by 5, so 10 is a multiple of 5.

✔ To generate all the factors of a number, begin by writing down 1, then leave some space, and then write the number itself. All remaining factors fall between these two numbers, so now check easy numbers like 2, 3, 4, and so forth, to see whether they're factors as well. For example, to generate the factors of 20, begin by writing down 1, then leaving some space, and then writing down 20. Now, include the factor 2 (and 10), and then 4 (and 5):

1   2   4   5   10   20

✔ To find the greatest common factor (GCF) for a set of numbers, generate the factors for every number in that set and pick the largest number that appears in every list of factors.

✔ Generating the multiples of a number is simple: Just multiply that number by 1, 2, 3, and so forth. For example, the first five multiples of 7 are 7, 14, 21, 28, and 35.

✔ To find the least common multiple (LCM) for a set of numbers, generate multiples of each number in that set and pick the smallest number that appears in every list of multiples.

Identifying Factors

211–225

211. Which of the following numbers have 2 as a factor?

i. 78

ii. 181

ii. 3,000

iv. 222,225

v. 1,234,569

212. Which of the following numbers have 5 as a factor?

i. 78

ii. 181

iii. 3,000

iv. 222,225

v. 1,234,569

213. Which of the following numbers have 3 as a factor?

i. 78

ii. 181

iii. 3,000

iv. 222,225

v. 1,234,569

214. Which of the following numbers have 10 as a factor?

i. 78

ii. 181

iii. 3,000

iv. 222,225

v. 1,234,569

215. Which of the following numbers have 7 as a factor?

i. 78

ii. 181

iii. 3,000

iv. 222,225

v. 1,234,569

216. How many factors does the number 12 have?

217. How many factors does the number 25 have?

218. How many factors does the number 32 have?

219. How many factors does the number 39 have?

220. How many factors does the number 41 have?

221. How many factors does the number 63 have?

222. How many factors does the number 90 have?

223. How many factors does the number 120 have?

224. How many factors does the number 171 have?

225. How many factors does the number 1,000 have?

Finding Nondistinct Prime Factors

226–232

226. How many nondistinct prime factors does 30 have?

227. How many nondistinct prime factors does 66 have?

228. How many nondistinct prime factors does 81 have?

229. How many nondistinct prime factors does 97 have?

230. How many nondistinct prime factors does 98 have?

231. How many nondistinct prime factors does 216 have?

232. How many nondistinct prime factors does 800 have?

Figuring the Greatest Common Factor

233–242

233. What is the greatest common factor (GCF) of 16 and 20?

234. What is the greatest common factor (GCF) of 12 and 30?

235. What is the greatest common factor (GCF) of 25 and 55?

236. What is the greatest common factor (GCF) of 26 and 78?

237. What is the greatest common factor (GCF) of 125 and 350?

238. What is the greatest common factor (GCF) of 28, 35, and 48?

239. What is the greatest common factor (GCF) of 18, 30, and 99?

240. What is the greatest common factor (GCF) of 33, 77, and 121?

241. What is the greatest common factor (GCF) of 40, 60, and 220?

242. What is the greatest common factor (GCF) of 90, 126, 180, and 990?

Mastering Multiples

243–249

243. How many multiples of 4 are between 1 and 30?

244. How many multiples of 6 are between 1 and 70?

245. How many multiples of 7 are between 1 and 100?

246. How many multiples of 12 are between 1 and 150?

247. How many multiples of 15 are between 1 and 175?

248. How many multiples of 16 are between 1 and 200?

249. How many multiples of 75 are between 1 and 1,000?

Looking for the Least Common Multiple

250–260

250. What is the least common multiple (LCM) of 6 and 8?

251. What is the least common multiple (LCM) of 7 and 11?

252. What is the least common multiple (LCM) of 4 and 14?

253. What is the least common multiple (LCM) of 12 and 15?

254. What is the least common multiple (LCM) of 8 and 18?

255. What is the least common multiple (LCM) of 20 and 45?

256. What is the least common multiple (LCM) of 8, 10, and 16?

257. What is the least common multiple (LCM) of 4, 12, and 18?

258. What is the least common multiple (LCM) of 3, 7, and 17?

259. What is the least common multiple (LCM) of 10, 14, and 24?

260. What is the least common multiple (LCM) of 11, 15, 16, and 25?

Chapter 8

Word Problems about Factors and Multiples

· ·

Factors and multiples both arise as a result of the operations of multiplication and division. Understanding how to work with factors and multiples becomes especially important as you move on to working with fractions in Chapter 9.

The Problems You'll Work On

In this chapter, the word problems that you face require you to do the following:

✔ Dividing people or objects into equal groups

✔ Finding a number that's divisible by a set of other numbers

✔ Generating the factors or prime factors of a number to solve a problem

✔ Finding the number of people in a group using multiples

✔ Using your knowledge of remainders to solve problems

What to Watch Out For

Here are a few tips to keep in mind as you tackle the word problems that follow:

✔ Read each question carefully to make sure you understand what it is asking.

✔ As you read, record and organize the information from the problem on scratch paper.

✔ Whenever possible, use the divisibility tricks from Chapter 6 in place of dividing.

✔ Use the techniques from Chapter 7 — generating factors, finding the prime factorization, and generating multiples — as needed.

Easy Word Problems

261–266

261. For a reading assignment, a group of children broke into groups of three children each. Later, for a math game, the same group divided themselves into groups of seven. In each case, no child was left out when groups were formed. If the class has fewer than 40 students, how many children does it include?

262. An animal shelter needed to transport 57 cats to their new facility. They used a quantity of large cages, placing the same number of cats in each cage. Assuming they placed fewer than seven cats in each cage, how many cats were transported in each cage?

263. A total of 91 people attended a wedding reception. They were seated at a set of tables, with each table accommodating the same number of guests. If there were more than eight tables, how many people sat at each table?

264. Mary Ann bought a bag of 105 pieces of candy. She divided it among her children so that each child received the same number of pieces, which was between 20 and 30 pieces of candy. How many children does Mary Ann have?

265. A parade committee wants to find the best way to arrange a marching band that will include 132 players. They want to march in a phalanx with a width of more than four but fewer than ten people, so that every row contains the same number of musicians. How many people should be in each row?

266. A group of 210 conference attendees were divided evenly into groups that contained more than 10 people but fewer than 20. What are the only two possible numbers of people that could have been in each group?

Intermediate Word Problems

267–274

267. Today, Maxine and Norma both completed the paperwork for routine inspections of their areas of a factory. Maxine is required to do the same inspection every 8 days and Maxine is required to do it every 14 days. How many days from now will be the next day on which they perform their respective inspections on the same day?

268. The size of a room is exactly 2,816 cubic feet. Each dimension of the room (length, width, and height) is a whole number of feet. If its height in feet is an odd number from 7 to 15, inclusive, what is the height of the room?

269. What is the smallest group of people that can be divided into subgroups of three people, four people, five people, or six people so that no people are left out of any group?

270. Marion tried to divide a basket of 100 apples among a group of friends. She found that two apples were left over after everyone had received an equal share. If the group of friends included fewer than 12 people, how many people were in the group?

271. What is the lowest square number that is divisible by both 3 and 4?

272. The area of a rectangular ballroom floor is exactly 168 square meters. Both sides of the room are a whole number of meters in length. If the longer side of the room is greater than 21 meters but less than 28 meters in length, how long is the shorter side of the room?

273. A manager divided a group of between 50 and 100 people into 21 teams, with each team containing the same number of people. Later, when she tried to arrange the same group of people into pairs, she found that one person was left over. How many people are in the manager's group?

274. What is the lowest number greater than 50 that is divisible by 7 but not by 2, 3, 4, 5, or 6?

Hard Word Problems

275–280

275. Andrea wanted to split a group of people into smaller subgroups that all contained the same number of people. She found that when she tried to split the group into subgroups containing two, three, or five people, there was always exactly one person left out. If the original group contained fewer than 50 people, how many people did it contain?

276. The number 1,260 is divisible by every number from 1 to 10 *except* which number?

277. Maxwell bought a package of between 70 and 80 colorful stickers to give to the children at his son's birthday party. He expected only nine children in total, so he divided the stickers up so that each child would have the same amount. But then another group of children arrived unexpectedly. Fortunately, he was able to divide the stickers up so that each child, again, received an equal number of stickers. Assuming the car contained fewer than nine additional children, how many children did it contain?

278. Before their graduation ceremony, a high school class sat in a set of seats that was arranged in a perfect square, with as many seats in each row as there were in each column. To receive their diplomas, they were instructed to walk to the stage in groups of eight. If the class had more than 200 students but fewer than 300, how many groups of eight students walked to the stage?

279. How many 2-digit numbers are also prime numbers?

280. When you divide a certain number by either 4 or 6, the remainder is 3. But when you divide the same number by 5, the remainder is 4. What is the lowest possible number that this could be?

Chapter 9

Fractions

● ●

*F*ractions are a common way of describing parts of a whole. They're commonly used for English weights and measures, especially for small measurements in cooking and carpentry.

The Problems You'll Work On

Here are the skills that you focus on in this chapter:

- ✔ Converting improper fractions to mixed numbers, and vice versa
- ✔ Increasing the terms of fractions and reducing fractions to lowest terms
- ✔ Cross-multiplying to compare the size of two fractions
- ✔ Applying the Big Four operations (adding, subtracting, multiplying, and dividing) to fractions and mixed numbers
- ✔ Simplifying complex fractions

What to Watch Out For

Here are a few things to remember as you begin to solve the problems in this chapter:

- ✔ Remember that the *numerator* of a fraction is the top number and the *denominator* is the bottom number.
- ✔ The *reciprocal* of a fraction is that fraction turned upside-down. For example, the reciprocal of $\frac{2}{3}$ is $\frac{3}{2}$.

Identifying Fractions

281–286

281. Identify the fraction of each circle that's shaded.

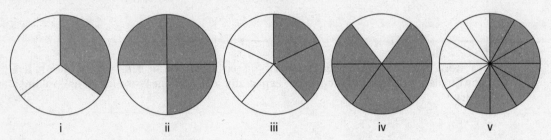

     i   ii   iii   iv   v

282. Identify the numerator and denominator of each fraction or number.

i. $\frac{1}{4}$

ii. $\frac{2}{9}$

iii. $\frac{9}{2}$

iv. 4

v. 0

283. Which of the following fractions are proper and which are improper?

i. $\frac{3}{8}$

ii. $\frac{5}{4}$

iii. $\frac{11}{12}$

iv. $\frac{1}{1,000}$

v. $\frac{101}{100}$

284. Change each of the following whole numbers to a fraction.

  i. 3

  ii. 10

  iii. 250

  iv. 2,000

  v. 0

285. Rewrite each of the following fractions as a whole number.

  i. $\frac{6}{2}$

  ii. $\frac{20}{5}$

  iii. $\frac{54}{6}$

  iv. $\frac{100}{50}$

  v. $\frac{150}{25}$

286. Find the reciprocal of the following numbers.

  i. $\frac{2}{7}$

  ii. $\frac{5}{3}$

  iii. $\frac{1}{10}$

  iv. 6

  v. $\frac{99}{100}$

Converting Numbers to Fractions

287–290: *Convert the following mixed numbers to improper fractions.*

287. $2\frac{1}{5}$

288. $4\frac{3}{7}$

289. $6\frac{1}{12}$

290. $9\frac{7}{10}$

Converting Fractions to Mixed Numbers

291–294: *Convert the following improper fractions to mixed numbers.*

291. $\frac{13}{3}$

292. $\frac{31}{2}$

293. $\frac{83}{5}$

294. $\frac{122}{11}$

Increasing Terms

295–300: *Increase the terms of the following fractions to the indicated amount.*

295. $\frac{3}{5} = \frac{?}{20}$

296. $\frac{2}{7} = \frac{?}{56}$

297. $\frac{4}{10} = \frac{?}{120}$

298. $\frac{9}{13} = \frac{?}{65}$

299. $\frac{6}{7} = \frac{?}{84}$

300. $\frac{13}{15} = \frac{?}{135}$

Reducing Terms

301. $\frac{8}{22}$

302. $\frac{18}{42}$

303. $\frac{15}{105}$

304. $\frac{270}{720}$

305. $\frac{375}{1,250}$

306. $\frac{138}{230} =$

Comparing Fractions

307. $\frac{5}{7} = \frac{8}{11}$?

308. $\frac{3}{10} = \frac{4}{13}$?

309. $\frac{2}{5} = \frac{5}{23}$?

310. $\frac{5}{12} = \frac{12}{29}$?

311. $\frac{8}{9} = \frac{104}{117}$?

312. $\frac{97}{101} = \frac{971}{1,002}$?

Multiplying and Dividing Fractions

313–322: *Multiply or divide the fractions. When needed, reduce all answers to lowest terms and express improper fractions as mixed numbers.*

313. $\frac{3}{5} \times \frac{4}{7} =$

314. $\frac{2}{15} \times \frac{5}{8} =$

315. $\frac{7}{12} \times \frac{9}{70} =$

316. $\frac{9}{11} \times \frac{15}{22} =$

317. $\frac{17}{39} \times \frac{13}{34} =$

318. $\frac{1}{6} \div \frac{5}{9} =$

319. $\frac{7}{8} \div \frac{2}{7} =$

320. $\frac{1}{40} \div \frac{2}{5} =$

321. $\frac{10}{13} \div \frac{20}{39} =$

322. $\frac{51}{55} \div \frac{17}{33} =$

Adding and Subtracting Fractions

323–328: *Add or subtract the fractions using the tricks shown at the beginning of this chapter. When needed, reduce all answers to lowest terms and express improper fractions as mixed numbers.*

323. $\frac{3}{7} + \frac{6}{7} =$

324. $\frac{5}{16} + \frac{9}{16} =$

325. $\frac{34}{35} - \frac{13}{35} =$

326. $\frac{1}{7} + \frac{1}{8} =$

327. $\frac{1}{6} - \frac{1}{15} =$

328. $\frac{1}{10} - \frac{1}{14} =$

Adding and Subtracting Fractions Using Cross-Multiplication

329–336: *Add or subtract the fractions using cross-multiplication. When needed, reduce all answers to lowest terms and express improper fractions as mixed numbers.*

329. $\frac{2}{5} + \frac{3}{7} =$

330. $\frac{3}{4} - \frac{1}{5} =$

331. $\frac{5}{8} + \frac{5}{6} =$

332. $\frac{5}{6} - \frac{1}{9} =$

333. $\frac{7}{15} + \frac{3}{20} =$

334. $\frac{2}{11} - \frac{1}{17} =$

335. $\frac{17}{20} - \frac{9}{50} =$

336. $\frac{1}{40} + \frac{44}{45} =$

Adding and Subtracting Fractions by Increasing Terms

337–342: *Add or subtract the fractions by increasing the terms of one fraction. When needed, reduce all answers to lowest terms and express improper fractions as mixed numbers.*

337. $\frac{3}{4} - \frac{1}{12} =$

338. $\frac{4}{5} + \frac{7}{20} =$

339. $\frac{3}{7} - \frac{5}{21} =$

340. $\frac{91}{100} + \frac{3}{20} =$

341. $\frac{12}{13} + \frac{40}{117} =$

342. $\frac{189}{190} - \frac{35}{38} =$

347. $\frac{3}{4} + \frac{2}{5} + \frac{9}{10} =$

348. $\frac{7}{10} + \frac{7}{12} - \frac{7}{15} =$

Adding and Subtracting Fractions by Finding a Common Denominator

343–348: Add or subtract the fractions by finding a common denominator. When needed, reduce all answers to lowest terms and express improper fractions as mixed numbers.

343. $\frac{5}{6} - \frac{2}{9} =$

344. $\frac{7}{8} + \frac{11}{12} =$

345. $\frac{3}{10} + \frac{19}{25} =$

346. $\frac{5}{6} - \frac{7}{15} =$

Multiplying and Dividing Mixed Numbers

349–352: Multiply or divide the mixed numbers. When needed, reduce all answers to lowest terms and express improper fractions as mixed numbers.

349. $1\frac{3}{4} \times 2\frac{1}{5} =$

350. $3\frac{2}{5} \times 1\frac{5}{6}$

351. $6\frac{2}{3} \div 4\frac{1}{6} =$

352. $10\frac{2}{3} \div 3\frac{1}{5} =$

Adding and Subtracting Mixed Numbers

353–362: *Add or subtract the mixed numbers. When needed, reduce all answers to lowest terms and express improper fractions as mixed numbers.*

353. $1\frac{1}{8} + 3\frac{3}{8} =$

354. $2\frac{5}{6} + 5\frac{5}{6} =$

355. $4\frac{3}{5} + 6\frac{1}{7} =$

356. $3\frac{1}{2} + 1\frac{5}{6} + 4\frac{3}{4} =$

357. $1\frac{1}{4} + 1\frac{3}{5} + 1\frac{7}{10} + 1\frac{19}{20} =$

358. $6\frac{7}{8} - 4\frac{1}{8} =$

359. $42\frac{5}{9} - 39\frac{1}{6} =$

360. $10\frac{11}{15} - 1\frac{2}{5} =$

361. $9\frac{1}{7} - 2\frac{6}{7} =$

362. $78\frac{3}{8} - 48\frac{5}{11} =$

Simplying Fractions

363–370: *Simplify the following complex fractions. When needed, reduce all answers to lowest terms and express improper fractions as mixed numbers.*

363. $\dfrac{\frac{1}{5} + \frac{2}{5}}{\frac{5}{6} - \frac{1}{6}} =$

364. $\dfrac{1 + \frac{3}{4}}{1 - \frac{1}{3}} =$

365. $\dfrac{2 - \frac{5}{8}}{\frac{3}{4} - \frac{1}{3}} =$

366. $\dfrac{\frac{1}{7}+\frac{1}{9}}{3-\frac{1}{3}} =$

367. $\dfrac{\frac{6}{7}-\frac{2}{9}}{\frac{1}{2}+\frac{5}{14}} =$

368. $\dfrac{13-\frac{1}{2}}{11+\frac{1}{9}} =$

369. $\dfrac{1+\frac{1}{6}+\frac{1}{9}}{3-\frac{1}{8}} =$

370. $\dfrac{1-\frac{1+\frac{2}{3}}{8}}{8-\frac{5}{1-\frac{1}{3}}} =$

Chapter 10

Decimals

· ·

Decimals are commonly used for money, as well as for weights and measures, especially when using the metric system. Generally speaking, decimals are easier to work with than fractions, provided that you know how to place the decimal point in your answer.

The Problems You'll Work On

In this chapter, you work on the following types of skills:

✔ Knowing how to change the most common decimals to fractions, and vice versa

✔ Calculating to convert between decimals and fractions

✔ Understanding repeating decimals and converting them to fractions

✔ Applying the Big Four operations (adding, subtracting, multiplying, and dividing) to decimals

What to Watch Out For

Here are a few tips for working with decimals:

✔ To change a decimal to a fraction, put the decimal in the numerator of a fraction with a denominator of 1. Then, continue to multiply both the numerator and denominator by 10 until the numerator is a whole number. If necessary, reduce the fraction. For example, change 0.62 to a fraction as follows:

$$0.62 = \frac{0.62}{1} = \frac{6.2}{10} = \frac{62}{100} = \frac{31}{50}$$

✔ To change a fraction to a decimal, divide the numerator by the denominator until the division either terminates or repeats.

✔ To change a repeating decimal to a fraction, put the repeating portion of the decimal (without the decimal point) into the numerator of a fraction. Use as a denominator a number composed only of 9s with the same number of digits as the numerator. If necessary, reduce the fraction. For example, change the repeating decimal $0.\overline{123}$ to a fraction as follows:

$$0.\overline{123} = \frac{123}{999} = \frac{41}{333}$$

✔ To add or subtract decimals, line up the decimal points.

✔ To multiply decimals, begin by multiplying without worrying about the decimal points. When you're done, count the number of digits to the right of the decimal point in each factor and add the result. Place the decimal point in your answer so that your answer has the same number of digits after the decimal point.

✔ To divide decimals, turn the *divisor* (the number you're dividing by) into a whole number by moving the decimal point all the way to the right. At the same time, move the decimal point in the *dividend* (the number you're dividing) the same number of places to the right. Then place a decimal point in the *quotient* (the answer) directly above where the decimal point now appears in the dividend.

✔ When dividing decimals, continue until the answer either terminates or repeats.

Converting Fractions and Decimals

371–394

371. Change each of the following decimals to fractions.

  i. 0.1

  ii. 0.2

  iii. 0.4

  iv. 0.5

  v. 0.6

372. Change each of the following decimals to fractions.

  i. 0.01

  ii. 0.05

  iii. 0.125

  iv. 0.25

  v. 0.75

373. What fraction is equal to 0.17?

374. What fraction is equal to 0.35?

375. What fraction is equal to 0.48?

376. What fraction is equal to 0.06?

377. What fraction is equal to 0.174?

378. What fraction is equal to 0.0008?

379. What fraction is equal to 6.07?

380. What fraction is equal to 2.0202?

381. What decimal is equal to $\frac{13}{100}$?

382. What decimal is equal to $\frac{143}{10,000}$?

383. What decimal is equal to $\frac{3}{20}$?

384. What decimal is equal to $\frac{6}{25}$?

385. What decimal is equal to $\frac{3}{8}$?

386. What decimal is equal to $\frac{9}{16}$?

387. What repeating decimal is equal to $\frac{1}{3}$?

388. What repeating decimal is equal to $\frac{4}{9}$?

389. What repeating decimal is equal to $\frac{1}{18}$?

390. What repeating decimal is equal to $\frac{123}{999}$?

391. What fraction is equal to the repeating decimal $0.\overline{6}$?

392. What fraction is equal to the repeating decimal 0.$\overline{7}$?

393. What fraction is equal to the repeating decimal 0.$\overline{81}$?

394. What fraction is equal to the repeating decimal 0.$\overline{497}$?

Adding and Subtracting Decimals

395–408

395. What is 3.4 + 0.76?

396. What is 821.7 + 0.039?

397. What is 2.35 + 66.1 + 0.7?

398. What is 912.4 + 60.278 + 31.92?

399. What is 81.222 + 5.4 + 0.098?

400. What is 745.21 + 8.88 + 0.6478 + 0.00295?

401. What is 0.982 + 3,381 + 0.009673 + 58,433.2 + 845.92?

402. What is 76.5 – 51.3?

403. What is 4.831 – 0.62?

404. What is 7.007 – 4.08?

405. What is 574.8 – 0.23?

406. What is 611 – 2.19?

407. What is $100 - 0.876$?

408. What is $20,304.007 - 1,147.0006$?

Multiplying and Dividing Decimals

409–420

409. What is 9.21×0.5?

410. What is 13.77×0.08?

411. What is 0.0734×9.2?

412. What is 1.098×5.07?

413. What is 287.4×0.272?

414. What is 0.014365×0.836?

415. What is $4.32 \div 0.6$?

416. What is $0.3 \div 0.008$?

417. What is $136.08 \div 0.021$?

418. What is $0.049 \div 1.6$?

419. What is $0.067 \div 3.3$?

420. What is $0.0001 \div 0.007$?

Chapter 11

Percents

● ●

Percents are commonly used in business to represent partial amounts of money. They're also used in statistics to indicate a portion of a data set. Percents are closely related to decimals, which means that they're easier to work with than fractions.

The Problems You'll Work On

Here are a few of the types of percent problems you find in this chapter:

- ✔ Converting between decimals and percents
- ✔ Changing numbers from percents to fractions, and vice versa
- ✔ Calculating basic percentages
- ✔ Solving more difficult percent problems by setting up word equations

What to Watch Out For

In this section, I provide some tips for dealing with the percent problems you find throughout this chapter.

- ✔ To change a percent to a decimal, move the decimal point two places to the left and drop the percent sign.
- ✔ To change a decimal to a percent, move the decimal point two places to the right and attach a percent sign.
- ✔ To change a percent to a fraction, drop the percent sign and put the number of the percent in the numerator of a fraction with a denominator of 100. If necessary, reduce the fraction.
- ✔ To change a fraction to a percent, first change the fraction to a decimal by dividing, as explained in Chapter 10. Then, change the decimal to a percent by moving the decimal point two places to the right and attaching a percent sign.
- ✔ Calculate simple percents by dividing. For example, to find 50% of a number, divide by 2; to find 25%, divide by 4; to find 20%, divide by 5; and so forth.

Converting Decimals, Fractions, and Percents

421–447

421. Find the decimal equivalent of each of these percents.

    i.   1%

    ii.   5%

    iii.   10%

    iv.   50%

    v.   100%

422. Find the percent equivalent of each of these decimals.

    i.   2.0

    ii.   0.20

    iii.   0.02

    iv.   0.25

    v.   0.75

423. Find the fractional equivalent of each of these percents.

    i. 10%

    ii. 20%

    iii. 30%

    iv. 40%

    v. 50%

424. Find the percent equivalent of each of these fractions.

    i. $\frac{1}{2}$

    ii. $\frac{1}{3}$

    iii. $\frac{2}{3}$

    iv. $\frac{1}{4}$

    v. $\frac{3}{4}$

425. What decimal is equivalent to 37%?

426. What is the decimal equivalence of 123%?

427. What decimal is equal to 0.08%?

428. What percent is equal to 0.77?

429. What percent is equivalent to 5.5?

430. What percent is the equivalent of 0.001?

431. What fraction is equivalent to 11%?

432. What fraction is equivalent to 65%?

433. What fraction is equivalent to 44%?

434. What is the fractional equivalent of 18.5%?

435. What fraction is equivalent to 650%?

436. To which fraction is 0.3% equivalent?

437. The percent 112.5% is equal to which fraction?

438. What fraction is equal to $83\frac{1}{3}\%$?

439. What percent is equal to the fraction $\frac{39}{50}$?

440. What percent is equal to the fraction $\frac{17}{20}$?

441. What percent is equal to the fraction $\frac{3}{8}$?

442. What percent is equal to the fraction $\frac{39}{40}$?

443. What percent is equal to the fraction $\frac{1}{12}$?

444. What percent is equal to the fraction $\frac{5}{11}$?

445. What percent is equal to the fraction $2\frac{3}{5}$?

446. What percent is equal to the fraction $\frac{777}{10,000}$?

447. What percent is equal to the fraction $1\frac{1}{1,000}$?

Solving Percent Problems

448–470

448. What is 50% of 20?

449. What is 25% of 60?

450. What is 20% of 200?

451. What is 10% of 130?

452. What is $33\frac{1}{3}$% of 99?

453. What is 1% of 2,400?

454. What is 18% of 50?

455. What is 32% of 25?

456. What is 12% of $33\frac{1}{3}$?

457. What's 8% of 43?

458. What's 41% of 17?

459. What's 215% of 3.2?

460. What's 7.5% of 10.8?

461. What percent of 40 is 30?

462. 20 is what percent of 160?

463. 72 is 25% of what number?

464. 85% of what number is 255?

465. 71% of what number is 6,035?

466. 108% of what number is 17,604?

467. What percent of 2.5 is 2.4?

468. 99.5 is 9.95% of what number?

469. $\frac{1}{2}$ is what percent of $\frac{1}{3}$?

470. $33\frac{1}{3}$ is 75% of what number?

Chapter 12

Ratios and Proportions

• •

A *ratio* is a comparison between two numbers. A *proportion* is a simple and useful equation based on a ratio. Ratios and proportions are closely related to fractions. In most cases, changing a ratio into a fraction will help you apply your knowledge of fractions to solve the problem you're facing.

The Problems You'll Work On

The following are a few examples of the types of problems you face in this chapter:

- ✔ Knowing how ratios and fractions are related
- ✔ Using ratios to set up a proportional equation
- ✔ Solving word problems using proportions

What to Watch Out For

Here are a few tips for sorting out problems that involve ratios and proportions:

- ✔ Ratios are easier to work with as fractions. Change a ratio to a fraction by putting the first number in the numerator and the second number in the denominator. For example, the ratio 2:5 is equivalent to the fraction $\frac{2}{5}$.

- ✔ A *proportion* is an equation that includes two ratios set equal to each other. For example:

$$\frac{\text{cars}}{\text{bicycles}} = \frac{2}{5}$$

- ✔ You can often solve a ratio problem by setting up a proportion. For example, suppose you have a (rich) friend who owns 20 bicycles and has the same proportion of cars to bicycles that you do. To find out how many cars she has, plug in the number 20 for bicycles in the preceding equation:

$$\frac{\text{cars}}{20} = \frac{2}{5}$$

- ✔ Now, multiply both sides of the equation by 20 to get rid of both fractions:

$$\frac{\text{cars}}{20} \times 20 = \frac{2}{5} \times 20$$

The result is the answer to the problem:

$$\text{cars} = \frac{2}{5} \times 20 = 8$$

Fractions and Ratios

471–483

471. If a family has four dogs and six cats, what is the ratio of dogs to cats?

472. If a club has 12 boys and 15 girls, what is the ratio of boys to girls?

473. If a room contains 42 people who are married and 30 people who are single, what is the ratio of married people to single people?

474. Karina earned $32,000 last year and Tamara earned $42,000. What is the ratio of Karina's earnings to Tamara's?

475. A runner ran 4.9 miles yesterday and 7.7 miles today. What is the ratio of the distance she ran yesterday to the distance she ran today?

476. Joe committed to volunteering a certain number of hours for his church every month. He fulfilled 1/5 of his monthly commitment on Saturday and 1/3 of it on Sunday. What is the ratio of the number of hours he volunteered Saturday to the number he volunteered on Sunday?

477. If a company has 10 management-level staff and 25 nonmanagement level staff, what is the ratio of managers to the entire staff of the company?

478. If a class has 10 sophomores, 12 juniors, and 8 seniors, what is the ratio of sophomores to juniors to seniors?

479. If a class has 10 sophomores, 12 juniors, and 8 seniors, what is the ratio of seniors to the entire class?

480. If a class has 10 sophomores, 12 juniors, and 8 seniors, what is the of juniors to the combined number of sophomores and seniors?

481. The first, second, and third floors of an apartment building have, respectively, 5, 7, and 6 residents. If one person moves from the first floor up to the second floor, what will be the resulting ratio of first-floor residents to second-floor residents to third-floor residents?

482. Ann decided to save money by replacing the incandescent light bulbs in her house with fluorescent ones. Originally, she was using 2,400 watts in total, but after replacing the bulbs, she reduced her usage by 1,800 watts. What is the proportion of her usage before and after changing the bulbs?

483. A building that stands 450 feet tall has a television tower on top that is an additional 75 feet. What is the ratio of the height of the building without the tower to the height of the building with the tower?

Using Equations to Solve Ratios and Proportions

484–500

484. A political organization has a 7:1 ratio of members who are registered to vote to members who aren't registered. If the organization has 28 registered members, how many nonregistered members does it have?

485. A house has a 9:2 ratio of windows to doors. If it has four doors, how many windows does it have?

486. Suppose that a store expects a 3 to 10 ratio of people who make a purchase to the number of people who enter the store. If 120 customers entered the store on a busy Saturday, how many made purchases?

487. A diet requires a 6:4:1 ratio of protein to fat to carbohydrates. If it permits 660 calories of daily fat intake, how many calories does it permit altogether?

488. A project manager estimates that her newest project will require a 2:9 ratio of team leaders to programmers. If she brings a total of 77 people into the project, how many of these will be team leaders?

489. A diner has an 8:5 ratio of dinner customers to lunch customers. If it averages 40 lunch customers, what is its average number of customers for both lunch and dinner?

490. A bookmobile has a 15 to 4 ratio of nonfiction books to fiction books. If it has 900 nonfiction books, how many books does it have altogether?

491. An organization has a 5:3:2 ratio of members from, respectively, Massachusetts, Vermont, and New Hampshire. If 60 members are from New Hampshire, how many are from Massachusetts?

492. An organization has a 5:3:2 ratio of members from, respectively, Massachusetts, Vermont, and New Hampshire. If 42 of its members are from Vermont, how many members are from either of the other two states?

493. An organization has a 5:3:2 ratio of members from, respectively, Massachusetts, Vermont, and New Hampshire. If the organization has a total of 240 members, how many are from Vermont?

494. Jason can swim 9 laps in the time it takes his cousin Anton to swim 5 laps. If the two boys swam a combined total of 140 laps in the same time span, how many of those laps did Jason swim?

495. A company has a 6 to 1 ratio of domestic to foreign sales revenue. It its total revenue last year was $350,000, how much of that was from foreign sales?

496. A restaurant sells a 5 to 3 ratio of red wine to white wine. If it sells 14 more bottles of red wine than white wine in one night, how many bottles does it sell altogether?

497. A portfolio made a 6% return on investment last year. What is the ratio of funds at the start of the year to the funds at the end of the year?

498. Before a recent vacation to Zurich, Karl traded $500 for 450 Swiss francs. When he returned to the United States, he still had 54 Swiss francs in his pocket. If he was fortunate enough to receive the same exchange rate, how many dollars did he get back?

499. Charles recently tracked his monthly spending and found that he spends 20% of his income on rent and 15% on transportation. If $3,250 goes to neither rent nor transportation, what is his rent each month?

500. In an alternative universe, multiplication is treated differently. For example:

$$\frac{1}{2} \times 3 = 2$$

Assuming that the product of multiplication in this other universe is proportional to ours, how would you solve the following equation in that universe:

$$\frac{1}{4} \times 12 = ?$$

Chapter 13

Word Problems for Fractions, Decimals, and Percents

....

Word problems provide you an opportunity to apply your skills to real-world problems. The problems in this chapter are divided into three main sections. Additionally, the section that includes percent problems also includes some very common problems in percent increase and percent decrease. In this chapter, you wrestle with a variety of word problems that require you to calculate using fractions, decimals, and percents.

The Problems You'll Work On

Here is a breakdown of the types of problems you solve in this chapter:

- ✔ Applying your understanding of fractions to solve word problems
- ✔ Solving word problems that involve decimals
- ✔ Using word equations to solve percent word problems
- ✔ Finding the answer to tricky percent increase and percent decrease problems

What to Watch Out For

The key to solving word problems is finding a way to set up each problem in a way that allows you to apply your skills at calculation. Here are a few tips to help you set up and solve the word problems you find in this chapter:

- ✔ For simpler problems, first find out which Big Four operation (adding, subtracting, multiplying, or dividing) you'll need to solve the problem.
- ✔ Follow the rules for operations with fractions, decimals, or percents as shown in Chapters 9, 10, and 11.
- ✔ Solve percent increase problems by adding the amount of the percent increase to 100%. For example, a 5% increase to an amount is equivalent to 105% of that amount.
- ✔ Solve percent decrease problems by subtracting the amount of the percent decrease *from* 100%. For example, a 10% decrease to an amount is equivalent to 90% of that amount.

Fraction Problems

501–520

501. Daniel is selling boxes of candy so that his soccer team can buy new uniforms. His uncle bought $\frac{1}{8}$ of the boxes he is selling and his aunt bought $\frac{1}{6}$ of them. What fraction of the total did his uncle and aunt buy altogether?

502. Jennifer ran $\frac{3}{5}$ of a mile and Luann ran $\frac{1}{2}$ of a mile. How much farther did Jennifer run than Luann?

503. What is one-fifth of $\frac{2}{3}$?

504. If you own $\frac{3}{5}$ of an acre of land and you subdivide it into four equal parts, what fraction of an acre is each part?

505. If the distance to and from your school is $1\frac{3}{8}$ miles, what is the distance one way?

506. If three children split 14 cookies evenly, how many cookies does each child get?

507. What is $\frac{1}{2}$ of $\frac{1}{4}$ of $\frac{1}{8}$ of $\frac{1}{16}$?

508. On a car trip, Arnold drove $\frac{1}{5}$ of the distance before he needed a break. Then, his wife Marion took over and drove $\frac{1}{3}$ of the total distance. At that point, what fraction of the total distance was still left for them to drive?

509. Jake practices basketball every day after school, from Monday through Friday, for $1\frac{1}{2}$ hours each day. On Saturdays and Sundays, he practices for $2\frac{1}{4}$ hours each day. How much does he practice every week?

510. An extra-large pizza was cut into 16 slices. Jeff took $\frac{1}{4}$ of the pizza, then Molly took 2 slices, and then Tracy took exactly half of the slices that were left. How many slices were left after Jeff, Molly and Tracy took their pieces?

511. Sylvia walked $2\frac{1}{2}$ miles on Friday, $3\frac{1}{4}$ on Saturday, and $4\frac{3}{10}$ on Sunday. How far did she walk during all three days combined?

512. Esther needs $12\frac{1}{2}$ feet of wood to build a set of shelves. She found $3\frac{1}{4}$ feet in her basement and another $4\frac{1}{2}$ feet in her garage. How many more feet does she need to buy?

513. Nate's mother bought a gallon of milk. He drank $\frac{1}{4}$ of the gallon on Monday, then drank $\frac{1}{4}$ of what was left on Tuesday. How much of the gallon was left after that?

514. A recipe for chocolate chip cookies requires $1\frac{1}{4}$ pounds of butter to make a batch of 25 cookies. How much butter do you need if you want make 150 cookies?

515. Theresa divided a jar containing $1\frac{1}{2}$ gallons of iced tea evenly into five glasses for a group of children. Then, she found out that one of the children didn't want iced tea, so she redistributed the iced tea into four glasses. How much extra iced tea did each of the four children receive?

516. Harry wrote a 650-word article in $3\frac{1}{4}$ hours. At the same rate, how long would he take to write a 750-word article?

517. Craig and his mom baked an apple pie and a blueberry pie. They cut the apple pie into four equal pieces and Craig ate one piece. Then, they cut the blueberry pie into six equal pieces, and Craig's mom ate a piece. How much pie was left over altogether?

518. David bought a cake for himself and his friends. He cut a piece for himself that was $\frac{1}{6}$ of the total cake. Then Sharon cut a piece that was $\frac{1}{5}$ of what was left. Then Armand cut a piece that was $\frac{1}{2}$ of what was left. How much of the cake was left after all three friends had taken their pieces?

519. Jared ran $\frac{3}{4}$ of a mile in 6 minutes. How many miles per hour is that?

520. Here's a tricky one: If $1\frac{1}{2}$ chickens can lay $1\frac{1}{2}$ eggs in $1\frac{1}{2}$ days, how many eggs can $3\frac{1}{2}$ chickens lay in 3 days?

Decimal Problems

521–535

521. Connie bought 2.7 kilos of chocolate in one store, 4.9 kilos in another store, and 3.6 kilos in a third store. She then split everything she bought evenly with a friend. How much chocolate did Connie end up with?

522. Blair is 0.97 meters tall and his father is 1.84 meters tall. How much taller than Blair is Blair's father?

523. Lauren measured her steps and found that each step she takes is 0.7 meters. Then she walked the length of her school in 87 steps. What is the length of her school in meters?

524. A water tank with a capacity of 861 gallons fills at a rate of 10.5 gallons per second. How long does it take to fill the tank?

525. On a recent vacation, Ed ran three times to a lighthouse, which was a distance of 3.4 miles each time. His wife, Heather, ran five times around a lake, which was a distance of 2.3 miles each time. How much farther did Heather run than Ed?

526. Myra's car has a gas tank with a 12.4-gallon capacity. She recently took a trip where she traveled 403 miles on one tank of gas. Assuming that her tank was empty at the end of the trip, how many miles per gallon did she get?

527. James read 111 pages in an hour. At that rate, how many pages could he read in 1 minute?

528. When making soup for a soup kitchen, Britney uses $3\frac{1}{2}$ large cans of soup stock. Each can contains 1.3 liters of stock. How many liters of soup stock does she use altogether?

529. Tony bought a car whose sticker price was $10,995. His car payment was $356.10 per month for 36 months. How much interest did he pay on the car over and above the sticker price?

530. Ronaldo ran the 100-yard dash three times in 12.6, 12.3, and 13.1 seconds. His friend Keith ran it in 11.8, 12.4, and 12.6 seconds. How much faster was Keith's total time as compared with Ronaldo's?

531. Emily paid a total of $187.50 to rent a car for three days. Her sister, Dora, used the car for half a day. How much should Dora pay Emily for the use of the car?

532. Stephanie paid $129 for a pool pass last summer. If she had paid on a daily basis, she would have had to pay $6.50 every time she entered. If she went to the pool 29 times over the course of the summer, how much did Stephanie save by buying the pool pass?

533. The cost of an adult ticket for a theme park is $57.60. Children between 6 and 12 years old pay half of the adult price, and children under 6 pay only one-third of the adult price. What is the total cost for 2 adults, 3 children between 6 and 12, and 2 children under 6?

534. In 1973, Secretariat ran the 1.5-mile Belmont Stakes in 2 minutes and 24 seconds. How many miles per hour did he average during this race?

535. Anita swam 0.8 miles on Monday. On Tuesday and Wednesday she increased her distance by a factor of 0.25 from the previous day. How far did she swim altogether in the three days?

Percent Problems

536–560

536. Angela spent 15 hours last week studying for a final exam in history. She spent 40% of this time working with a set of flash cards she had made. How much time did she spend working with her flash cards?

537. An ad for a laptop claims that the laptop weighs 10% less than its main competitor. If the competitor weighs 1.1 kilos, how much does the laptop weigh?

538. Randy is on a diet that allows him to eat 2,000 calories per day. He wants to eat a piece of cake that's 700 calories. What percent of his daily total would that be?

539. Beth recently received a raise, bringing her pay from $11.50 to $13.80 per hour. What percent raise did she receive?

540. Geoff completed 35% of an 850-mile car trip on the first day. How far did he drive that day?

541. Over the weekend, Nora read 55% of a 420-page book. How many pages did she read?

542. Kenneth mowed the lawn 25 times last year from May to September, with 52% of this work occurring in May and June. How many times did he mow the lawn from July to September?

543. If 32.5% of a 60-minute television show is commercials, how many minutes of commercials are there during this hour?

544. Before a big party, Jason spent 3 hours and 45 minutes cleaning his apartment. He worked for 45 minutes cleaning the windows. What percentage of his cleaning time did he spend on the windows?

545. Eve has saved $8,000 for college. Fifteen hundred of this came from a scholarship through a local youth group. What percentage of Eve's savings came from this scholarship?

546. Janey set a goal of practicing the violin for 400 hours over her summer vacation. She has already practiced for 290 hours. What percentage of her goal has she completed?

547. In preparation for a new work assignment in Florence, Stephen spent 45 hours in an intensive class in Italian. This represented 15% of his preparation for the job. How many hours of preparation did Steven receive altogether?

548. A building has a ground-floor atrium that is 6.25 meters tall. This atrium is 5% of the total height of the building. What is the building's height?

549. Karan spends 28% of her monthly income on her mortgage. Her mortgage payment is $1,736. What is her monthly income?

550. As an associate in a law firm, Madeleine earns $135,000 a year. This is 225% of her previous earnings before attending law school. What was her salary in her previous job?

551. If you invest $12,000 and earn a 10% return on your investment, how much money do you have?

552. A television that usually sells for $750 is on sale for 15% off. What is its sale price?

553. Suppose you buy dinner in a restaurant, and the check comes to $26.00. If you wish to leave an 18% tip, rounded to the nearest whole dollar amount, how much will you pay for dinner?

554. The listed price for a house is $229,000. If the owner reduces this price by 3% and then rounds off to the nearest $1,000, what is the new price?

555. If you wish to leave at least a 15% tip for a lunch that costs $8.20, what is the least amount you can leave?

556. The yearly interest on a loan is 14.5%. If you borrow $4,250, what is the total that you must pay back, assuming that you pay the loan in one year?

557. In a biology experiment, Judy found that the mass of a yeast culture increased by 7.5%. If the original mass of the culture was 3 grams, what was it at the end of the experiment?

558. Marian bought a car priced at $18,000. She talked the dealer into giving her a 9% discount, but then he was required to charge her 8% of the discounted price in sales tax. How much did Marian end up paying for the car?

559. Dane invested $7,200 and, after a loss, had only $6,624 left. By what percent did his investment decrease?

560. At a restaurant, Beth left an 18% tip for her server, paying a total price of $32.45. What was the price of the check without the tip?

Chapter 14

Scientific Notation

· ·

Scientific notation is an alternative to standard notation (that is, the way you normally write numbers). Standard notation can be awkward for writing very large and very small numbers, such as 19,740,000,000,000 and 0.0000000000291. Scientific notation allows you to express very large and very small numbers in a more compact way. For example: $19,740,000,000,000 = 1.974 \times 10^{13}$ and $0.0000000000291 = 2.91 \times 10^{-11}$.

The Problems You'll Work On

Here are some of the scientific notation problems you'll see in this chapter:

- ✔ Converting numbers from standard notation to scientific notation
- ✔ Converting numbers from scientific notation to standard notation
- ✔ Multiplying numbers in scientific notation

What to Watch Out For

Keep these tips in mind as you work through the questions:

- ✔ A number in scientific notation always includes two parts: A *decimal* no less than 1.0 but less than 10, multiplied by a *power of 10*.

- ✔ To convert a large number from standard notation to scientific notation, begin by multiplying the number by 10^0, then move the decimal point one place to the left and add 1 to the exponent until the decimal portion of the number is between 1 and 10.

- ✔ To convert a small number from standard notation to scientific notation, begin by multiplying the number by 10^0, then move the decimal point one place to the right and subtract 1 from the exponent until the decimal portion of the number is between 1 and 10.

- ✔ To convert a large number from scientific notation to standard notation, move the decimal point one place to the right and subtract 1 from the exponent until the exponent reduces to 0; then drop the power of 10.

- ✔ To convert a small number from scientific notation to standard notation, move the decimal point one place to the left and add 1 to the exponent until the exponent increases to 0; then drop the power of 10.

- ✔ To multiply two numbers in scientific notation, multiply the two decimals and add the two exponents. If the resulting decimal is greater than 10, move the decimal point one place to the left and add 1 to the exponent.

Converting Standard Notation and Scientific Notation

561–580

561. What is 1,776 in scientific notation?

562. Express 900,800 in scientific notation.

563. What is 881.99 in scientific notation?

564. What is the equivalent of 987,654,321 in scientific notation?

565. What is ten million in scientific notation?

566. How do you translate 0.41 into scientific notation?

567. What is 0.000259 equal to in scientific notation?

568. What is the equivalent of 0.001 in scientific notation?

569. What is the value of 0.0000009 in scientific notation?

570. How do you represent one-millionth in scientific notation?

571. What is the equivalent of 2.4×10^3 in standard notation?

572. What is the equivalent of 3.45×10^5 in standard notation?

573. The distance from the earth to the sun is approximately 1.5×10^8 kilometers. How many kilometers is that in standard notation?

574. Scientists estimate that the universe is approximately 1.46×10^{10} years old. How many years is that in words?

575. A parsec is a unit of astronomical distance approximately equivalent to 3.1×10^{13} miles. How many miles is this in standard notation?

576. What is the equivalent of 7.5×10^{-2} in standard notation?

577. What is the equivalent of 3×10^{-3} in words?

578. An inch is approximately equivalent to 2.54×10^{-5} kilometers. What is this number equivalent to in standard notation?

579. What is the value of 8×10^{-10} in standard notation?

580. What is the value of 1×10^{-7} in words?

Multiplying Numbers in Scientific Notation

581–590

581. What do you get when you multiply 2×10^3 by 3×10^4?

582. What is 1.1×10^6 multiplied by 7×10^2?

583. What is the result when you multiply 1.6×10^9 by 4.2×10^1?

584. What is 3.5×10^{-4} times 2.51×10^7?

585. What do you get when you multiply 2.5×10^{-3} by 4.9×10^4?

586. What is 1.9×10^{15} times 8×10^{-27}?

587. What is $\left(6.7 \times 10^{1}\right) \times \left(5.1 \times 10^{-1}\right)$?

588. What is $\left(3.29 \times 10^{20}\right) \times \left(7.7 \times 10^{0}\right)$?

589. What is $\left(2.23 \times 10^{7}\right) \times \left(4.67 \times 10^{-9}\right) \times \left(7.13 \times 10^{8}\right)$?

590. What is $\left(9 \times 10^{-16}\right) \times \left(4.7 \times 10^{-24}\right) \times \left(8.2 \times 10^{45}\right)$?

Chapter 15

Weights and Measures

· ·

The two measurement systems used most frequently throughout the world are the English and metric systems. Both systems provide units for measuring length, volume, weight, time, and temperature. The English system is used most frequently in the United States; the metric system is used throughout the rest of the world.

The Problems You'll Work On

In this chapter, use the formulas at the beginning of each section to answer the questions that follow. Most of these questions are practical problems that test your ability to convert from one type of unit to another.

- ✔ Changing units within the English system
- ✔ Making decimal conversions within the metric system
- ✔ Converting temperature between English and metric units
- ✔ Estimating conversions between English and metric units

What to Watch Out For

Here is some additional information about the English and metric systems of measurement:

- ✔ English units of length are inches, feet, yards, and miles; metric units are based on the meter.

- ✔ English units of weight are ounces, pounds, and tons; metric units of mass (similar to weight) are based on the gram.

- ✔ English units of volume are fluid ounces, cups, pints, quarts, and gallons; metric units are based on the liter.

- ✔ English units of temperature are degrees Fahrenheit; metric units are degrees Celsius (also called Centigrade).

- ✔ Both measurement systems measure time in terms of seconds, minutes, hours, days, weeks, and years.

English Measurements

591–603 *Use the following information about English measurement units.*

1 foot = 12 inches

1 yard = 3 feet

1 mile = 5,280 feet

1 pound = 16 ounces

1 ton = 2,000 pounds

1 gallon = 4 quarts

1 quart = 2 pints = 4 cups

1 cup = 8 fluid ounces

1 year = 365 days

1 week = 7 days

1 day = 24 hours

1 hour = 60 minutes

1 minute = 60 seconds

591. How many inches are in 13 feet?

592. How many minutes are in 18 hours?

593. How many ounces are in 15 pounds?

594. How many quarts are in 55 gallons?

595. How many inches are in 3 miles?

596. How many ounces are in 13 tons?

597. How many seconds are in a week?

598. How many fluid ounces are in 17 gallons?

599. A marathon is 26.2 miles. If you estimate that each step a runner takes is 1 yard long, how many steps are in a marathon?

600. The St. Louis Arch weighs 5,199 tons. How many ounces is that?

601. An average life span is approximately 80 years. How many seconds are in an average life span? (Assume that 1 year = 365 days and ignore the additional days that leap years would add.)

602. A raindrop is approximately $\frac{1}{90}$ fluid ounces. How many raindrops are in a gallon?

603. The width of a football field is $53\frac{1}{3}$ yards. How many widths of a football field equal a mile?

Metric Units

604–614 *Use the following information about the metric system:*

Nano = 0.000000001 (One billionth)

Micro = 0.000001 (One millionth)

Milli = 0.001 (One thousandth)

No prefix = 1 (One)

Kilo = 1,000 (One thousand)

Mega = 1,000,000 (One million)

Giga = 1,000,000,000 (One billion)

Tera = 1,000,000,000,000 (One trillion)

1 meter = 100 centimeters

604. How many milliliters are in 25 liters?

605. How many tons are in 800 megatons?

606. How many nanoseconds are in 30 seconds?

607. How many centimeters are in 12 kilometers?

608. How many milligrams are in 17 megagrams?

609. How many kilowatts are in 900 gigawatts?

610. How many microdynes are in 88 megadynes?

611. How many millimeters are in 333 terameters?

612. Water flows over Niagara Falls at a rate of 567,811 liters per second. How many liters flow over the falls in a microsecond?

613. If a nanogram contains 1,000 picograms and a petagram contains 1,000 teragrams, how many picograms are in a petagram?

614. If a computer can download 5 kilobytes of information in a nanosecond, how many terabytes of information can be downloaded in 1 second?

Temperature Conversions

615–620 *Convert between Celsius and Fahrenheit degrees using these formulas:*

$$C = (F - 32) \div 1.8$$
$$F = (C \times 1.8) + 32$$

615. In Celsius, water freezes at 0°C and boils at 100°C. The midpoint between these two temperatures is 50°C. What is the equivalent of this midpoint temperature in degrees Fahrenheit?

616. A healthy body temperature is considered to be 98.6°F. What is the equivalent in Celsius?

617. The average person considers 72 degrees Fahrenheit to be the most comfortable temperature. What is this temperature in degrees Celsius, to the nearest whole degree?

618. The hottest weather ever recorded on Earth was 136 degrees Fahrenheit. What is the equivalent temperature in degrees Celsius, to the nearest whole degree?

619. The melting point of iron is 1,535 degrees Celsius. What is this temperature in degrees Fahrenheit?

620. Absolute zero is the lowest possible temperature, the point at which all molecular motion stops. In Celsius, it is −273.15°C. What is the equivalent temperature in degrees Fahrenheit?

Converting English and Metric Units

621–630 *Use the following approximations for converting between English and metric units:*

1 meter ≈ 3 feet = 1 yard

1 kilometer ≈ $\frac{1}{2}$ mile

1 liter ≈ 1 quart = $\frac{1}{4}$ gallon

1 kilogram ≈ 2 pounds

621. Approximately how many miles is 20 kilometers?

622. Approximately how many liters are in 12 gallons?

623. If a man weighs 180 pounds, approximately how many kilograms does he weigh?

624. The tallest building in the world is the Burj Khalifa, which stands approximately 828 meters tall. What is its approximate height in feet?

625. If a tree is 60 feet tall, what is its approximate height in meters?

626. If an elephant weighs 5,000 kilograms, approximately how many pounds does it weigh?

627. If you run 5 miles a day every day for 2 weeks, approximately how many kilometers have you run?

628. If a commuter puts 95 liters of gasoline into her car every week, approximately how many gallons of gasoline does she use in 4 weeks?

629. If the length of a swimming pool is about $\frac{1}{32}$ mile, what's its approximate length in meters?

630. If a machine is calibrated to deliver exactly 40 milliliters of saline solution, approximately how many fluid ounces does it deliver?

Chapter 16

Geometry

● ●

Geometry is the study of points, lines, angles, shapes on the plane, and solids in space. In this chapter, you hone your geometry skills with a variety of problems that ask you to calculate the measurement of angles, shapes, and solids.

The Problems You'll Work On

Here's a list of the types of problems you work on in this chapter:

✔ Measuring angles

✔ Finding the area and perimeter of squares and rectangles

✔ Calculating the area of parallelograms and trapezoids

✔ Knowing the formulas for the area and circumference of circles

✔ Using the area formula for triangles

✔ Working with right triangles using the Pythagorean theorem

✔ Finding the volume of some common solids

What to Watch Out For

The following information will be useful to you as you work on the problems in this chapter:

✔ You need to know basic geometric formulas to find the area and perimeter of squares and rectangles, and the area of parallelograms, trapezoids, and triangles.

✔ You need to know the formulas to find the diameter, circumference, and area of circles.

✔ You need to know the formulas to find the volume of cubes, rectangular solids, cylinders, spheres, pyramids, and cones.

✔ You should be familiar with the Pythagorean theorem as well as the formulas associated with right triangles.

Angles

631–645

631. Find the value of *n*.

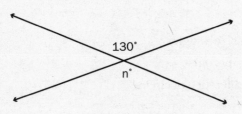

632. Find the value of *n*.

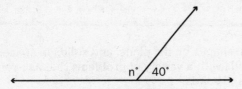

633. Find the value of *n*.

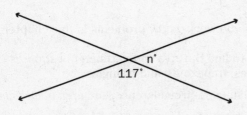

634. Find the value of *n*.

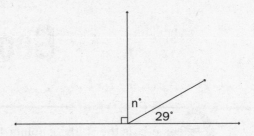

635. *ABCD* is a square. Find the value of *n*.

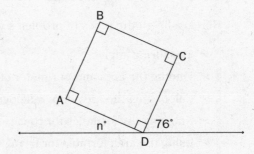

636. Find the value of *n*.

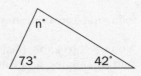

637. Find the value of *n*.

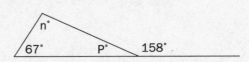

638. *ABC* is a right triangle. Find the value of *n*.

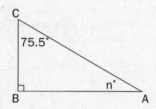

639. *ABCD* is a rectangle. Find the value of *n*.

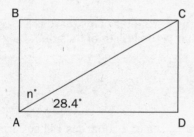

640. Find the value of *n*.

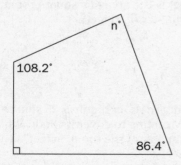

641. Find the value of *n*.

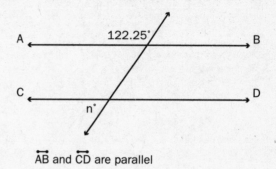

$\overleftrightarrow{AB}$ and $\overleftrightarrow{CD}$ are parallel

642. Find the value of *n*.

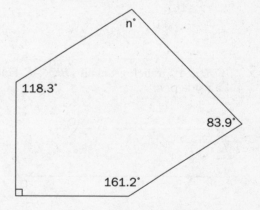

643. *ABC* is isosceles. Find the value of *n*.

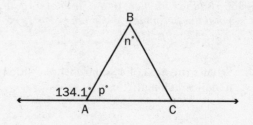

644. *AC* is a diameter of the circle. Find the value of *n*.

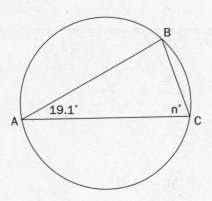

645. *BCDE* is a parallelogram and *BE* = *AE*. Find the value of *n*.

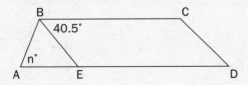

Squares

646–655 *Use the formulas for the area of a square (A = s²) and the perimeter of a square (P = 4s) to answer the questions.*

646. What's the area of a square whose side is 6 inches in length?

647. What's the perimeter of a square that has a side that's 7 meters in length?

648. What's the area of a square with a side that's 101 miles long?

649. If the side of a square is 3.4 centimeters, what is its perimeter?

650. If the perimeter of a square is 84 feet long, what is the length of its side?

651. If the area of a square is 144 square feet, what is its perimeter?

652. What is the area of a square room that has a perimeter of 62 feet?

653. A square room requires 25 square yards of carpeting to cover its floor. What is the perimeter of the room, in feet? (1 yard = 3 feet)

654. If each side of a square field is exactly 3 miles, what is its area in square feet? (1 mile = 5,280 feet)

655. The perimeter of a square park, in kilometers, is 10 times greater than its area in square kilometers. What is the length of one side of this park?

Rectangles

656–665 *Use the formulas for the area of a rectangle (A = lw) and the perimeter of a rectangle (A = 2l + 2w) to answer the questions.*

656. What is the area of a rectangle with a length of 8 centimeters and a width of 3 centimeters?

657. If a rectangle has a length of 16 meters and a width of 2 meters, what's its perimeter?

658. If the length of a rectangle is 4.3 feet and its width is 2.7 feet, what is its area?

659. What is the perimeter of a rectangle whose length is $\frac{7}{8}$ inch and whose width is $\frac{3}{4}$ inch?

660. What is the area of the rectangle below?

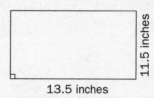

11.5 inches

13.5 inches

661. What is the area of the rectangle below?

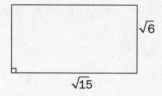

$\sqrt{6}$

$\sqrt{15}$

662. If the area of a rectangle is 100 square feet and the width is 5 feet, what is its perimeter?

663. If the area of a rectangle is 30 square inches and its length is 8 inches, what is its perimeter?

664. A rectangular picture frame has a length of 2 feet. If the area of the picture in the frame is 156 square inches, what is the perimeter of the frame?

665. If the perimeter of a rectangle is 54 and its area is 72, what is the length of the rectangle. (Hint: The length and width are both whole numbers.)

Parallelograms and Trapezoids

666–675 *Use the formulas for the area of a parallelogram (A = bh) and the area of a trapezoid $(A = \frac{b_1 + b_2}{2} h)$ to answer the questions.*

666. What's the area of the parallelogram below?

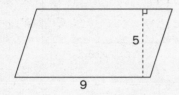

667. What's the area of the parallelogram below?

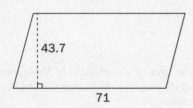

668. What's the area of the parallelogram below?

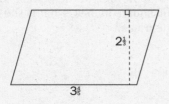

669. What's the area of the trapezoid below?

670. What's the area of the trapezoid below?

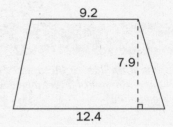

671. What's the area of the trapezoid below?

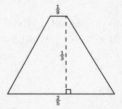

672. If the area of a parallelogram is 94.5 square centimeters and its base is 7 centimeters, what is its height?

673. What is the height of a trapezoid that has an area of 180 and bases of lengths 9 and 21?

674. Suppose a parallelogram has an area of $\frac{4}{9}$ and a height of $\frac{5}{7}$. What is the length of its base?

675. If a trapezoid has an area of 45, a height of 3, and one base of length 4.5, what is the length of the other base?

Area of Triangles

676–685 Use the formula for the area of a triangle ($A = \frac{1}{2}bh$) to answer the questions.

676. What is the area of a triangle with a base of 9 inches and a height of 8 inches?

677. If a triangle has a base that's 3 meters long and a height that's 23 meters in length, what is its area?

678. A triangle has a base of length $\frac{4}{9}$ and a height of length $\frac{3}{8}$. What is its area?

679. What is the area of the triangle below?

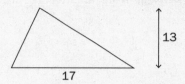

680. What is the area of the triangle below?

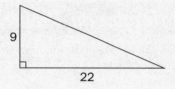

681. A right triangle has two legs of length 4 centimeters and 12 centimeters. What is its area?

682. What is the base of a triangle with an area of 60 square meters and a height of 4 meters?

683. What is the height of a triangle with an area of 78 square inches and a base that's 1 foot long?

684. What is the height of a triangle with a base of $\frac{5}{7}$ in length and an area of $\frac{5}{8}$ in²?

685. If the area of a triangle is 84.5 and the height and base are both the same length, what is the height of the triangle?

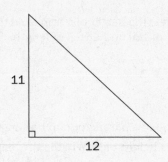

The Pythagorean Theorem

686–695 *Use the Pythagorean theorem $(a^2 + b^2 = c^2)$ to answer the questions.*

686. If the two legs of a right triangle are 3 feet and 4 feet, what is the length of the hypotenuse?

687. What is the length of the hypotenuse of a right triangle whose two legs are 10 centimeters and 24 centimeters?

688. If a right triangle has two legs of length 4 and 8, what is the length of its hypotenuse?

689. What is the length of the hypotenuse of the following triangle?

690. What is the length of the hypotenuse of the triangle below?

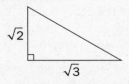

691. What is the length of the hypotenuse of the triangle below?

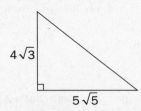

692. If a right triangle has two legs of lengths $\frac{5}{13}$ and $\frac{12}{13}$, what is the length of the hypotenuse?

693. If a right triangle has two legs of lengths $\frac{1}{3}$ and $\frac{1}{4}$, what is the length of the hypotenuse?

694. What is the length of the shorter leg of the triangle below?

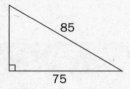

695. What is the length of the longer leg of the triangle below?

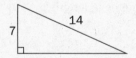

Circles

696–709 Use the formulas for the diameter of a circle $(D = 2r)$, the area of a circle $(A = \pi r^2)$, and the circumference of a circle $(C = 2\pi r)$ to answer the questions.

696. What's the diameter of a circle with a radius of 8?

697. What's the area of a circle with a radius of 11?

698. If a circle has a radius of 20, what's the length of its circumference?

699. What's the area of the circle below?

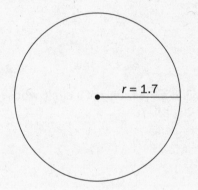

700. What's the circumference of the circle below?

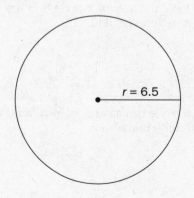

701. If a circle has a diameter of 99, what's its circumference?

702. What's the circumference of a circle whose diameter is $\frac{5}{6}$?

703. If a circle has a diameter of 100, what is its area?

704. What's the radius of a circle whose area equals 81π?

705. If a circle has a circumference of 66π, what is its radius?

706. What's the area of a circle that has a circumference of 10.8π?

707. If a circle has an area of $\frac{4}{25}\pi$, what is its circumference?

708. What's the radius of a circle whose area is 16?

709. What's the area of a circle whose circumference is 18.5?

Volume

710–730 *Use the solid geometry formulas:*

Volume of a cube: $V = s^3$

Volume of a box: $V = lwh$

Volume of a cylinder: $V = \pi r^2 h$

Volume of a sphere: $V = \frac{4}{3}\pi r^3$

Volume of a pyramid (with a square base): $V = \frac{1}{3}s^2 h$

Volume of a cone: $V = \frac{1}{3}\pi r^2 h$

710. What's the volume of a cube that has a side of 12 inches in length?

711. What is the volume of the cube below?

7.5

712. If a cube has a volume of 1,000,000 cubic inches, what is the length of its side?

713. What is the volume of a box that's 15 inches long, 4 inches wide, and 10 inches tall?

714. If a box is 8.5 inches in width, 11 inches in length, and 3.5 inches in height, what's its volume?

715. What is the volume of a box whose three dimensions are $\frac{1}{4}$ inch, $\frac{5}{8}$ inch, and $\frac{11}{16}$ inch?

716. Suppose a box with a volume of 20,000 cubic centimeters has a length of 80 centimeters and a width of 50 centimeters. What is the height of the box?

717. If a box has a volume of 45.6 cubic inches, a length of 10 inches, and a height of 100 inches, what is its width?

718. If a cylinder has a radius of 2 feet and a height of 6 feet, what's its volume?

719. Suppose a cylinder has a radius of 45 and a height of 110. What is its volume?

720. What's the volume of a cylinder whose radius is 0.4 meter and whose height is 1.1 meters?

721. A cylinder has a radius of $\frac{3}{4}$ inch and a height of $\frac{5}{8}$ inch. What's its volume?

722. Suppose a cylinder has a volume of 58.5π cubic feet and a radius of 3 feet. What is its height?

723. What is the volume of a sphere with a radius of 3 centimeters?

724. If a sphere has a radius of $\frac{3}{8}$ inch, what's its volume?

725. What is the volume of a sphere that has a radius of 1.2 meters?

726. Suppose a sphere has a volume of $\frac{1}{6}\pi$ cubic feet. What is its radius?

727. A pyramid has a square base whose side has a length of 4 inches. If its height is 6 inches, what is the volume of the pyramid?

728. Suppose a pyramid with a square base has a volume of 80 cubic meters and a height of 15 meters. What is the length of the side of its base?

729. What is the volume of a cone that's 10 inches high and whose circular base has a radius of 30 inches?

730. If a cone has a volume of 132π and a radius of 6, what is its height?

Chapter 17

Graphing

. . . .

A *graph* is a visual representation of mathematical data. Graphs make it easy to see how a set of values relate to each other. In this chapter, you hone your graph-reading skills. You also get some practice working with the most ubiquitous graph in mathematics, the *xy*-graph.

The Problems You'll Work On

Here's a preview of the types of problems you'll solve in this chapter:

- ✔ Working with graphs that display data: bar graphs, pie charts, line graphs, pictographs
- ✔ Plotting points, calculating slope, and finding the distance between two points on the *xy*-graph

What to Watch Out For

The problems in this chapter provide practice in working with a variety of types of graphs. Here are descriptions of each type of graph, with some additional pointers for answering questions about the *xy*-graph.

- ✔ A *bar graph* allows you to compare values that are independent from each other.
- ✔ A *pie chart* provides a picture of how an amount is divided into percentages.
- ✔ A *line graph* shows you how a value changes over time.
- ✔ A *pictograph*, similar to a bar graph, allows you to compare values that are independent from each other.
- ✔ An *xy graph* (also called a *Cartesian graph*) allows you to plot points as pairs of values (x, y).
- ✔ Find the slope of a line that passes between two points on the *xy*-graph by starting with the point on the left and proceeding toward the point on the right. First count the number of points *up* or *down* and then count the number of points *over* (that is, from left to right), and make a fraction from these two numbers.
- ✔ Find the distance between two points on the *xy*-graph by drawing a right triangle using the distance between these two points as the hypotenuse. Calculate the length of this hypotenuse using the Pythagorean theorem: $a^2 + b^2 = c^2$.

Bar Graph

731-736 *The bar graph shows the number of dollars that each of six people collected for charity during their office walk-a-thon.*

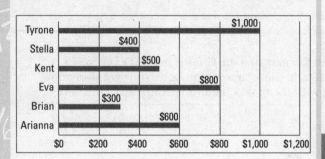

731. Which person collected exactly $200 more than Brian?

732. How much money did the three women (Arianna, Eva, and Stella) collect altogether?

733. What fraction of the total amount did Stella collect?

734. What is the ratio of Stella's total to Tyrone's?

735. If Eva had collected $300 less, which other person would have collected the same amount of money as she did?

736. To the nearest whole percentage point, what percentage of the total did Arianna and Tyrone contribute together?

Pie Chart

737-742 *The pie chart shows the percentages of time that Kaitlin devotes to studying for her five college classes.*

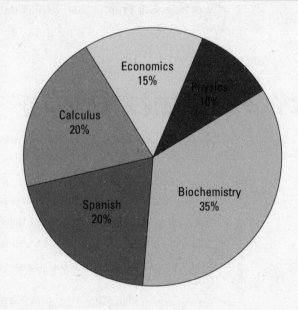

737. Which *two* classes combined take up exactly half of Kaitlin's time?

738. Which *three* classes combined take up exactly 55% of Kaitlin's time?

739. If Kaitlin spent 20 hours last week studying, how much time did she spend studying for her Spanish class?

740. If Kaitlin spent 1.5 hours more studying for her Calculus class than her Economics class, how much time did she spend studying altogether?

741. If Kaitlin spent three hours last week studying for her Physics class, how many hours did she spend studying altogether?

742. If Kaitlin spent two hours last week studying for her Economics class, how much time did she spend studying for her Biochemistry class?

Line Graph

743–747 *The line graph shows the monthly net profit statement for Amy's Antiques from January to December.*

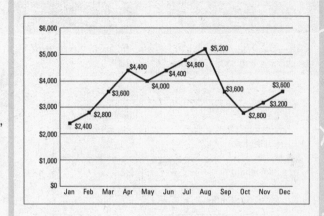

743. In which month was the net profit the same as the net profit in February?

744. What is the total net profit for the first quarter of the year (that is, January, February, and March combined)?

745. Which month shows the same increase in net profit, when compared with the previous month, as was shown in April when compared with March?

746. Which pair of consecutive months shows a combined net profit of exactly $8,800?

747. Which month accounts for approximately 5% of the total yearly net profit?

Population Pictograph

748–752 *The pictograph below shows the population of the six towns in Alabaster County.*

City	Population
Barker Lake	𝙭𝙭𝙭𝙭𝙭
Jamesburg	𝙭𝙭𝙭𝙭𝙭𝙭
Morrissey Station	𝙭𝙭𝙭𝙭𝙭𝙭𝙭𝙭
Plattfield	𝙭𝙭𝙭𝙭𝙭𝙭𝙭𝙭𝙭𝙭𝙭
Ravenstown	𝙭𝙭𝙭
Talkingham	𝙭𝙭𝙭𝙭𝙭𝙭𝙭

1 stick figure = 2,000 people

748. What is the population of the largest town in Alabaster County?

749. Which town contains slightly more than $\frac{1}{6}$ of the population of the county?

750. To the nearest whole percentage, what percent of the county lives in Morrissey Station?

751. Which pair of towns has a combined population that's 1,000 greater than Plattfield?

752. Imagine that the population of Talkingham increased by the equivalent of one stick figure in the graph and the other five towns remained constant in their populations. In that case, to the nearest whole percent, what percentage of Alabaster county would reside in Talkingham?

Pie Chart

753–757 *The pie chart below shows the election turnout for the mayor of Branchport.*

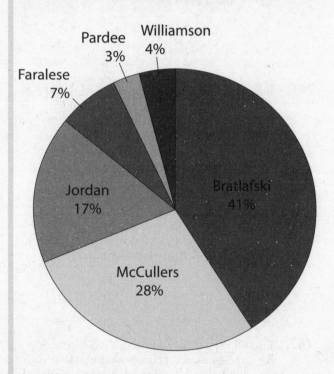

753. Together, the two top candidates received what percentage of the vote?

754. Which pair of candidates received 35% of the vote together?

755. If 100,000 votes were cast, how many more votes did Faralese receive than Williamson?

756. If Jordan received 34,000 votes, how many votes were cast altogether?

757. If Bratlaski received 53,200 more votes than Pardee, how many votes did McCullers get?

Trees Pictograph

758–762 *The pictograph shows the number of trees planted in six counties.*

City	
Birmingham County	🌲🌲🌲🌲🌲
Calais County	🌲🌲
Dublin County	🌲🌲🌲🌲🌲🌲🌲🌲
Edinburgh County	🌲🌲🌲
London County	🌲🌲🌲🌲🌲
Manchester County	🌲🌲🌲🌲🌲🌲

1 tree figure = 250 trees

758. How many trees were planted altogether in Edinburgh and Manchester Counties?

759. How many trees were planted altogether in the six counties?

760. Which two counties together account for 50% of the trees planted among the six counties?

761. Which county accounts for 18.75% of the trees planted among the six counties?

762. Imagine that 1,000 more trees had been planted in Calais County and all the other counties remained the same. In that case, what fraction of the total number of trees would have been planted in Calais County?

Cartesian Graph

763–770 *Use the* xy *graph to answer the questions.*

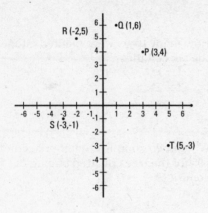

763. Name the point at each of the following coordinates:

    i.   (1, 6)

    ii. (–3, –1)

    iii. (–2, 5)

    iv. (3, 4)

    v.   (5, –3)

764. What is the slope of the line that passes through both the origin (0, 0) and Q?

765. What is the slope of the line that passes through both the origin (0, 0) and S?

766. What is the slope of the line that passes through both P and Q?

767. What is the slope of the line that passes through both R and T?

768. What is the slope of the line that passes through both S and T?

769. What is the distance between the origin (0, 0) and P?

770. What is the distance between R and S?

Chapter 18

Statistics and Probability

· ·

*S*tatistics is the mathematics of real-world events. In statistics, you analyze *data sets* — information measured from actual events — using a variety of tools. *Probability* measures the likelihood of an event whose outcome is unknown. Calculating probability rests upon ordered methods of counting possible outcomes of events. In this section, you solve problems in both statistics and probability.

The Problems You'll Work On

Here are the main skills you practice in the following problems.

- ✔ Finding the mean, median, and mode of a data set
- ✔ Calculating a weighted mean
- ✔ Counting independent and dependent events
- ✔ Deciding the probability of an event

What to Watch Out For

Here are some tips for calculating statistics and probability in the problems that follow:

- ✔ When a question asks for the average without specifying which type, calculate the mean using the following formula: $Mean = \dfrac{Sum\ of\ items}{Number\ of\ items}$.
- ✔ To find the median, rank all values in the data set from least to greatest and find the middle number; that's the median. If there are two middle numbers (that is, if the data set has an even number of values) calculate the median as the mean of these two values.
- ✔ To count the number of possible outcomes for independent events, multiply the number of possible outcomes in each case. For example, when you roll a pair of dice, each die has six possible outcomes, so there are $6 \times 6 = 36$ possible outcomes altogether.
- ✔ To count the number of possible outcomes for dependent events, multiply the number of possible outcomes in each case, taking into account any previous outcomes. For example, when you draw two letters from a bag of five letters, you can draw any of five letters first and any of the remaining four letters second, so there are $5 \times 4 = 20$ possible outcomes altogether.
- ✔ Calculate probability using the following formula: $Probability = \dfrac{Target\ outcomes}{Total\ outcomes}$.

Finding Means

771–782 Use the formula for the mean $(Mean = \frac{Sum\ of\ items}{Number\ of\ items})$ to answer the questions.

771. What is the mean of 4, 9, and 11?

772. What is the mean of 2, 2, 16, 29, and 81?

773. What is the mean of 245, 1,024, and 2,964?

774. What is the mean of 17, 23, 35, 64, and 102?

775. What is the mean of 3.5, 4.1, 9.2, and 19.6?

776. What is the mean of 7.214, 91.8, and 823.24?

777. What is the mean of $\frac{1}{5}$ and $\frac{1}{9}$?

778. What is the mean of $3\frac{1}{3}$, $4\frac{1}{5}$, and $6\frac{1}{2}$?

779. Kathi earned $40 on Monday and $75 on Tuesday. She also worked on Wednesday, and found that the mean of her earnings from Monday through Wednesday was $60. How much did she earn on Wednesday?

780. Antoine went on a camping trip where he hiked for four days. On the first three days, he hiked 8 miles, 4.5 miles, and 6.5 miles. At the end of the four days, he found that he had hiked an average of 7 miles per day. How many miles did Antoine hike on the last day?

781. For a science class, Marie is measuring how far a caterpillar can crawl. She measured its distance for 5 minutes and found that it traveled an average of $5\frac{1}{8}$ inches. If it crawled a total of $18\frac{3}{8}$ in the first four minutes, how many inches did it travel in the last minute?

782. Eleanor studied an average of 9 hours per day on the 7 days leading up to her Bar Exam. If she studied only 4 hours on the final day, but an average of 11 hours per day on the three days preceding this, what was her average daily study time on the first three days of the week?

Finding Weighted Means

783–791 *Find the weighted mean.*

783. Four of the five senior homeroom classes at Metro High School have 16 students and one has 21 students. What is the average class size among these five homerooms?

784. At a political event, five speakers gave 8-minute speeches and three speakers gave 10-minute speeches. What was the average speech length for the event?

785. Jake earned $280 per week for the first four weeks of his summer vacation. Then he got a raise and earned $340 per week for the remaining six weeks. What was his average weekly income over the summer?

786. If you save $1,000 per month for six months, then $500 per month for the next four months, and then $700 per month for the next two months, what will be your average monthly savings for the year, to the nearest whole dollar?

787. Angela ran 4 laps around a track in 31 minutes and 50 seconds, then ran another 6 laps in 48 minutes and 40 seconds. What was her average time per lap for the 10 laps?

788. Kevin's math teacher gave 12 quizzes this semester, each worth a maximum of 10 points. Kevin scored 10 points on 2 tests, 9 points on 5 tests, 8 points on 3 tests, 7 points on 1 test, and 6 points on 1 test. What was his average grade on the tests?

789. The first floor of a 20-story building is 20 feet in height. The next 4 floors are each 12 feet, and the remaining 15 floors are each 8 feet. What is the average floor height?

790. Elise loves jigsaw puzzles. She completed two 300-piece puzzles in three days, then completed three 1,000-piece puzzles in a week, and then completed four 500-piece puzzles in six days. On average, how many puzzle pieces did she put together on each day?

791. On a long car trip, Gerald drove for 45 minutes at 75 miles per hour, then for an hour and a half at 65 miles per hour, then for 75 minutes at 55 miles per hour, and finally for 1 hour at 70 miles per hour. What was his average speed for the whole trip?

Medians and Modes

792–796 *Calculate the median or the mode (or both) of the given data set to answer the question.*

792. What is the median of the following data set: 8, 14, 14, 15, 19, 21, and 23?

793. What is the median of the following data set: 17, 24, 37, 45, 48, and 70?

794. What is the mode of the following data set: 8, 13, 13, 15, 16, 16, 16, 29, and 33?

795. What is the difference between the median and the mode of the following data set: 2, 3, 3, 4, 5, 5, 5, 6, 7, 9, 10, 12, 15, and 15?

796. Which integer from 11 to 15 is *not* the mean, the median, or a mode of the following data set: 1, 1, 11, 11, 11, 12, 13, 14, 14, 14, and 63?

Independent Events

797–803 *Calculate the number of possible independent events.*

797. How many different ways can you roll a pair of six-sided dice?

798. How many different ways can you roll an 8-sided die, a 12-sided die, and a 20-sided die?

799. Jeff brought two suits, four shirts, and seven ties with him on a business trip. How many different ways could he pick out one suit, one shirt, and one tie to wear?

800. A complementary breakfast includes a choice among four types of eggs, three types of breakfast meat, two types of potatoes, and four types of beverages. How many different breakfast combinations are possible?

801. A survey includes 10 yes or no questions. How many different ways can the survey be filled out?

802. A monogram is a set of three initials. How many monograms are possible using the 26 letters in the English alphabet, A through Z?

803. A computer password is exactly four symbols, each of which can be either a digit (from 0 to 9) or a letter (from A to Z). How many different passwords are possible?

Dependent Events

804–816 *Calculate the number of dependent events.*

804. A bag contains four letter tiles with the letters A, B, C, and D. In how many different orders can you pull these four letters out of the bag?

805. Five friends arrive at a restaurant one at a time. In how many different possible orders can these five people arrive?

806. If you are making a pizza with six toppings — sausage, pepperoni, onions, mushrooms, green peppers, and garlic — in how many different orders can you place these six toppings on the pizza one at a time?

807. A summer reading list includes eight books, which you can choose to read in any order. In how many different orders can you choose to read the eight books?

808. Twenty children are playing a game that requires a pitcher, a catcher, and a runner. How many such permutations are possible among the 20 children?

809. A monogram is a set of three initials. How many monograms *in which no letter is repeated* are possible using the 26 letters A through Z?

810. A magic trick requires you to pick 3 cards from a deck of 52 cards and keep them in the order you picked them. How many different ways can you do this?

811. A club that has 16 members elects a president, a vice-president, a treasurer, and a secretary. No club member can hold more than one elected position. How many different ways can these four positions be filled?

812. How many different five-digit numbers contain no repeated digits? (Note that a number cannot have 0 as its first digit.)

813. How many different ways can you arrange the letters in the word CAPSIZE so that the first letter is a vowel (A, E, or I)?

814. How many different ways can you arrange the letters in the word CAPSIZE so the first two letters are both vowels (A, E, or I) and the third and fourth letters are both consonants (C, P, S, or Z)?

815. Three women and three men all arrived one at a time to a job interview. If all three women arrived before all three men, in how many different orders could the six people have arrived?

816. Three women and three men all arrived one at a time to a job interview. If each man arrived just after a woman, in how many different orders could the six people have arrived?

Probability

822–835 *Calculate probability using the probability formula* Probability = $\dfrac{Target\ outcomes}{Total\ outcomes}$ *to answer the question.*

817. A bag contains ten tickets printed with the numbers 1 through 10. If you pull one ticket from the bag at random, what is the probability that it will be the number 1?

818. A bag contains ten tickets printed with the numbers 1 through 10. If you pull one ticket from the bag at random, what is the probability that it will be an even number?

819. A bag contains ten tickets printed with the numbers 1 through 10. If you pull one ticket from the bag at random, what is the probability that it will be a number greater than 6?

820. A bag contains ten tickets printed with the numbers 1 through 10. If you pull two tickets from the bag at random, what is the probability that they will both be odd numbers?

821. What's the probability of rolling the number 2 on a six-sided die?

822. What's the probability of rolling a number greater than 2 on a six-sided die?

823. What's the probability of rolling a number *other* than 2 on a six-sided die?

824. If you roll a pair of six-sided dice, what's the probability of rolling a pair of numbers that adds up to 12?

825. If you roll a pair of six-sided dice, what's the probability of rolling a pair of numbers that adds up to 10?

826. If you roll a pair of six-sided dice, what's the probability of rolling a pair of numbers that adds up to either 7 or 11?

827. If you roll three six-sided dice, what is the probability that they will total 16?

828. If you pick a card from a deck of 52 cards, what is the probability that it will be one of the four aces in the deck?

829. If you pick a card from a deck of 52 cards, what is the probability that it will be one of the 13 hearts in the deck?

830. In a deck of cards, a "court card" is one of the four kings, four queens, or four jacks. If you pick a card from a deck of 52 cards, what is the probability that it will be a court card?

831. If you pick 2 cards from a deck of 52 cards, what is the probability that both of them will be aces?

832. If you pick 4 cards from a deck of 52 cards, what is the probability that all of them will be aces?

833. Three women and three men all arrived one at a time to a job interview. Assuming that they arrived in a random order, what is the probability that the first person to arrive was a woman?

834. Three women and three men all arrived one at a time to a job interview. Assuming that they arrived in a random order, what is the probability that the first three people to arrive were all women?

835. Three women and three men all arrived one at a time to a job interview. Assuming that they arrived in a random order, what is the probability that each man arrived immediately after a different woman?

Chapter 19

Set Theory

Set theory is, not surprisingly, the study of sets — that is, collections of things. Every member of a set is called an *element* of that set; for example, the set {1, 4, 5} has three elements. In this chapter, you solve problems that strengthen your skills in basic set theory.

The Problems You'll Work On

The problems in this chapter focus on the following skills:

- ✔ Performing the operations of union, intersection, and relative complement on sets
- ✔ Working with sets of numbers such as even and odd numbers, positive and negative numbers, and so forth
- ✔ Finding the complement of a set using the set of integers as the universal set
- ✔ Solving problems using Venn diagrams

What to Watch Out For

Here is a list of the basic operations and other concepts you'll need to solve the problems in this chapter:

- ✔ The *union* ($\cup$) of two sets includes every element that's in *either* set.
- ✔ The *intersection* ($\cap$) includes every element that's in *both* sets.
- ✔ The *relative complement* (–) includes every element of the first set that's *not* an element of the second.
- ✔ The empty set ($\varnothing$) contains no elements.
- ✔ The *complement* of a set is every element of the universal set that's not an element of the set.
- ✔ When solving problems using Venn diagrams, to find the *total* number of elements in a set, add the two numbers in the circle representing that set.
- ✔ When solving problems using a Venn diagram, the numbers written in the diagram are (from left to right) the number of elements in
    - The first set but not the second set
    - Both sets
    - The second set but not the first set
    - Neither set

Performing Operations on Sets

836–847 *Set P = {1, 3, 5, 7, 9}, Set Q = {6, 7, 8}, Set R = {1, 2, 4, 5}, and Set S = {3, 6, 9}.*

836. What is $P \cup Q$?

837. What is $P \cap Q$?

838. What is $P - Q$?

839. What is $Q - P$?

840. What is $Q \cup S$?

841. What is $R \cap S$?

842. What is $(P \cup Q) \cap R$?

843. What is $P \cup (Q \cap R)$?

844. What is $P - (Q \cup S)$?

845. What is $(P - Q) \cup S$?

846. What is $(Q - S) \cap R$?

847. What is $(Q \cup R) \cap (P - S)$?

Set Relationships

848–851

848. What is the intersection of the set of integers and the set of even integers?

849. What is the relative complement of the set of positive integers and the set of even integers?

850. What is the union of the set of odd negative integers and the set of even positive integers?

851. What is the intersection of the set of odd negative integers and the set of even positive integers?

Complements

852–855 *Use the set of integers {..., –2, –1, 0, 1, 2, ...} as the universal set.*

852. What is the complement of a set of integers greater than 7?

853. What is the complement of the set of odd integers?

854. What is the complement of ∅?

855. What is the complement of the set of nonnegative integers?

Venn Diagrams

856–860

856. The Venn diagram below shows the number of students in the Jefferson High School Hiking Club who are seniors, honors students, both, or neither.

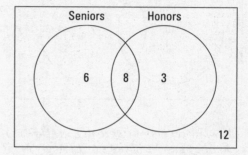

How many students are members of the club?

857. The Venn diagram below shows information about the number of people at the Kinney family reunion who are actually surnamed Kinney, who live out of state, both, or neither.

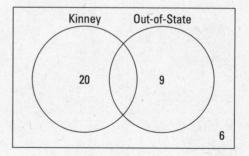

If 42 people attended the reunion, how many of these people are surnamed Kinney?

858. The Venn diagram below shows information about the number of people in a theater group who were in the cast of their last two plays. The cast of *12 Angry Men* included 13 people and the cast of *Long Day's Journey Into Night* included 5 people.

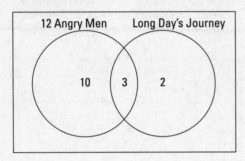

How many people were in *Long Day's Journey Into Night* but not *12 Angry Men*?

859. A board comprising 18 people includes 7 officers and 10 people who have served more than one term. If exactly 2 officers have served more than one term, how many people on the board are nonofficers who are serving their first term?

860. A teacher asked her class of 24 students how many owned at least one cat, and 15 raised their hands. Then she asked how many owned at least one dog, and 10 raised their hands. If three children in the class own neither a dog nor a cat, how many own at least one cat but no dog?

Chapter 20

Algebraic Expressions

• •

Algebra allows you to solve problems that are too difficult for arithmetic alone. In algebra, you begin to work with *variables* (such as x). Each variable in an algebra problem stands in for an unknown number. In this chapter, the focus is on algebraic expressions, which are the building blocks of the algebraic equations that you work with in Chapter 21.

The Problems You'll Work On

The questions here fall into three general categories:

✔ **Evaluation**: When you know the value of every variable in an algebraic expression, you can *evaluate* that expression by plugging in this value. For example, if $x = 4$, then $3x + 6 = 3(4) + 6 = 18$.

✔ **Simplification**: Even if you don't know the value of every variable in an algebraic expression, you can often *simplify* it. For example, $10y - 7y + 2x = 3y + 2x$.

✔ **Factoring**: In some cases, you can *factor* an algebraic expression by dividing every term by a common factor. For example, when you factor out 3 from the expression $3x - 6$, the result is $3(x - 2)$.

What to Watch Out For

Here is some additional terminology that you may find helpful:

✔ An *equation* is a mathematical statement with an equal sign, such as $2 + 2 = 4$. An *algebraic equation* includes at least one variable — for example, $3x - 6 = 10y - 7y + 2x$.

✔ Every equation includes two *expressions,* placed on opposite sides of the equal sign. For example, $2 + 2$, 4, $3x + 6$, and $10y$ are four examples of expressions. An *algebraic expression* includes at least one variable — for example, $3x - 6$ and $10y - 7y + 2x$.

✔ Expressions can be separated into *terms*. For example, the expression $3x - 6$ has two terms: $3x$ and -6. The expression $10y - 7y + 2x$ has three terms: $10y$, $-7y$, and $2x$. As you can see, each term carries the sign (+ or −) that precedes it.

✔ Terms are composed of *coefficients* (the number and sign) and *variables* (the letter or letters). For example, the term $3x$ comprises the coefficient 3 and the variable x. Similarly, the term $-7y$ comprises the coefficient -7 and the variable y. As you can see, each coefficient includes the sign (+ or −) that precedes it.

Evaluating

861–870

861. Evaluate $3x - 5y$, given that $x = 9$ and $y = 4$.

862. Evaluate $x^2 - 8y + 10$, given that $x = 5$ and $y = -2$.

863. Evaluate $-2x^2y + 11$, given that $x = -6$ and $y = -1$.

864. Evaluate $4x^3 + 5xy - y$, given that $x = -2$ and $y = 3$.

865. Evaluate $3x^2 + 5xy - 0.25$, given that $x = 0.5$ and $y = -0.5$.

866. Evaluate $5\left(x^2y^2 + x\right)^2$, given that $x = 0.1$ and $y = 3$.

867. Evaluate $0.7\left(0.1xy + 0.2x - 0.3y^2\right)$, given that $x = 7$ and $y = 9$.

868. Evaluate $\dfrac{x}{y} + \dfrac{3y}{4x}$, given that $x = 5$ and $y = -8$.

869. Evaluate $\dfrac{\left(10xy + y\right)^3}{2y}$, given that $x = -1$ and $y = 2$.

870. Evaluate $\left(\dfrac{x}{y}\right)^4\left(\dfrac{2y}{5x}\right)^2$, given that $x = -2$ and $y = 3$.

Simplifying

871–893

871. Simplify by combining like terms.
$2x + 3y + 5x - y$

872. Simplify by combining like terms.
$2x^3 + 3x^2 + 5x - 9 + 6x^3 - 10x^2 - 2x - 9$

873. Simplify by combining like terms.
$8x + 0.1 + 5y + 3xy + 5 - 0.1x - 0.3 + xy$

874. Simplify by combining like terms.
$x + \dfrac{1}{3}x^2 - \dfrac{1}{4}x^3 - \dfrac{2}{5}x + \dfrac{5}{8}x^3$

875. Simplify by multiplying. $(4x^3)(5x^4)$

876. Simplify by multiplying. $(2x^2y^2)(6xy^4)$

877. Simplify by multiplying. $(7x^2yz)(x^3y)(3yz^4)$

878. Simplify by applying the rule for simplifying exponents. $(9x^3)^2$

879. Simplify by applying the rule for simplifying exponents. $(6x^2y^4z^5)^3$

880. Simplify by applying the rule for simplifying exponents and then multiplying. $(4xy)^2(6x^2)(2y^4x)^3$

881. Simplify by dividing. $\dfrac{8x^4y^3}{4xy^2}$

882. Simplify by dividing. $\dfrac{(4x^5)^2}{(2x^2)^3}$

883. Simplify by dividing. $\dfrac{x(4xy^2)^3}{y^5(8x^2)^2}$

884. Simplify. $x+(3y-x-5)+8$

885. Simplify. $3y-(6x+4y-5)+(x-2)$

886. Simplify. $3x(6x+4y-8)-(9xy-11)$

887. Simplify. $-6xy(7+z)-yz(3x-4)$

888. Simplify. $x^2(6x+4)+3x(x+2)-9(x^2-7)$

889. Simplify by multiplying. $(x+3)(x-4)$

890. Simplify by multiplying. $(2x+1)(5x+3)$

891. Simplify by multiplying. $(x^2-2)(x+7)$

892. Simplify by multiplying. $4x(x-6)(x-10)$

893. Simplify by multiplying. $(x+1)(x-2)(x+4)$

Factoring

894–909

894. Factor. x^2-3x

895. Factor. x^5+x^2

896. Factor. $6x^8-x^7-4x^6$

897. Factor. $-12x^9+6x^6+4x^3$

898. Factor. $24x^{10}+15x^9+9x^4$

899. Factor. $x^2y^3+x^7y^7+x^4y$

900. Factor. $8x^{11}y^{14}+20x^9y^8-40x^6y^{10}$

901. Factor. $36xyz^5-24xy^8z^4+90x^6y^4z^3$

902. Factor. x^2-64

903. Factor. $9x^2-4$

904. Factor. $49x^2-100y^2$

905. Factor. $x^4 - 16y^{10}$

906. Factor. $x^2 + 9x + 14$

907. Factor. $x^2 - 11x + 18$

908. Factor. $x^2 + x - 20$

909. Factor. $x^2 - 10x - 24$

Simplifying by Factoring

910–920

910. Simplify by factoring.
$$\frac{x}{x^2 + x}$$

911. Simplify by factoring.
$$\frac{x^2 - x}{x - 1}$$

912. Simplify by factoring.
$$\frac{x^2 + x^4}{3 + 3x^2}$$

913. Simplify by factoring.
$$\frac{8x^5 + 4x^2}{10x^7 + 5x^4}$$

914. Simplify by factoring.
$$\frac{2x^3 + 6x^2 + 8x}{5x^2 + 15x + 20}$$

915. Simplify by factoring.
$$\frac{x^2 - 4}{x + 2}$$

916. Simplify by factoring.
$$\frac{x^2 - y^2}{2x + 2y}$$

917. Simplify by factoring.
$$\frac{4x^2 - 25}{8x + 20}$$

918. Simplify by factoring.

$$\frac{x-2}{16x^3-64x}$$

919. Simplify by factoring.

$$\frac{x^2-36}{x^2-7x+6}$$

920. Simplify by factoring.

$$\frac{x^2+12x+20}{x^2-x-6}$$

Chapter 21

Solving Algebraic Equations

• •

Algebra was invented to solve problems that could not be solved easily by arithmetic alone. Most of these types of problems can be stated as equations that contain at least one variable. Solving an equation means finding the value of the variable in that equation. This process often relies on your ability to work with algebraic expressions (see Chapter 20).

The Problems You'll Work On

Here are a few of the skills you'll be working on as you solve the problems in this chapter:

- ✔ Solving simple equations just by looking at them
- ✔ Solving equations by isolating the variable
- ✔ Removing parentheses from both sides of the equation before solving
- ✔ Solving equations that contain decimals and fractions
- ✔ Solving simple quadratic equations by factoring

What to Watch Out For

Here are a few guidelines to keep you on track while solving algebraic equations:

- ✔ Always keep equations balanced — any operation you perform on one side, you must perform on the other.

- ✔ Isolate all terms that contain the variable on one side of the equation and all constant terms on the other side, then combine terms on both sides and divide by the coefficient of the term with the variable.

- ✔ Remove parentheses from both sides of the equation before attempting to isolate the variable.

- ✔ Solve equations with decimals just as you would equations without them.

- ✔ When solving equations that contain fractions, remove the fractions first, either by cross-multiplication or by multiplying each term by a common denominator.

- ✔ Solve quadratic equations by factoring and then splitting the equation into two separate equations. For example, change $x^2 + 2x - 3 = 0$ to $(x - 1)(x + 2) = 0$, then split it into the equations $x - 1 = 0$ and $x + 2 = 0$. So the answer is $x = 1$ and $x = -2$.

Simple Equations

921–924

921. Solve the following five equations by just looking at them.

 i. $6 + x = 14$

 ii. $21 - x = 9$

 iii. $7x = 63$

 iv. $x \div 1 = 14$

 v. $99 \div x = 9$

922. Solve the following five equations by using inverse operations.

 i. $68 + x = 117$

 ii. $x - 83 = 29$

 iii. $13x = 585$

 iv. $41x = 3{,}116$

 v. $x \div 13 = 19$

923. Solve the equation $9x + 14 = 122$ by guessing and checking.

924. Solve the equation $30x + 115 = 985$ by guessing and checking.

Isolating Variables

925–933

925. $6x - 3 = 27$

926. $9n + 14 = 11n$

927. $v + 18 = -5v$

928. $9k = 3k + 2$

929. $2y + 7 = 3y - 2$

930. $m + 24 = 3m - 8$

931. $7a + 7 = 13a + 27$

932. $3h - 2h + 15 = 5h - 7h$

933. $6x + 4 + 2x = 3 + 9x - 10$

Solving Equations with Decimals

934–936

934. $2.3w + 7 = 3.7w$

935. $-1.9p + 7 = 2.1p - 7$

936. $0.8j - 2.4j + 1 = 9.4j + 0.7$

Solving Equations with Parentheses

937–945

937. $3 + (x - 1) = 2x - (5x + 7)$

938. $7u - (10 - 3u) = 5(3u + 4)$

939. $-(2k - 6) = 5(1 + 8k) + 5$

940. $6x(3 + 3x) + 39 = 9x(11 + 2x) - 3x$

941. $1.3(5v) = v + 11$

942. $0.2(15y + 2) = 0.5(8y - 3)$

943. $1.75(44m + 36) = 73m$

944. $1.8(3n - 5) = 3(4.3n + 5)$

945. $4.4(3s + 7) = 4s + 4(s - 0.2) - 13.5s - 5.8$

Solving Equations with Fractions

946–963

946. $\frac{1}{6}n = 7$

947. $\frac{2}{11}w = 8$

948. $\frac{3}{4}y = \frac{5}{9}$

949. $\frac{q}{7} = \frac{q-1}{9}$

950. $\frac{c+2}{5} = \frac{1-3c}{6}$

951. $\frac{t-10}{3t} = \frac{t}{3t+6}$

952. $\frac{z+1}{z-2} = \frac{z-1}{z+3}$

953. $\frac{3b-4}{6b} = \frac{2b-5}{4b+1}$

954. $p + \frac{p}{9} = 20$

955. $\frac{d}{2} + \frac{d}{3} = 5$

956. $\frac{3s}{2} + \frac{3s}{4} = s + 2$

957. $\frac{2r}{5} + \frac{r+1}{6} = r - 1$

958. $\frac{j}{2} + \frac{3}{4} = \frac{3j}{4}$

959. $\frac{1}{2} + \frac{5k}{8} = \frac{k}{4}$

960. $\frac{8a}{3} + \frac{1}{4} = \frac{a}{6}$

961. $\frac{h}{30} + \frac{3h}{20} = \frac{11}{10}$

962. $\frac{2}{3}k + 5 = \frac{1}{9}(2k + 1)$

963. $\frac{1}{2}(3 - y) + \frac{1}{4} = \frac{1}{8}(2y + 6)$

Factoring

964–970

964. What value or values other than $x = 0$ solve the equation $5x^4 = 35x^3$?

965. What value or values other than $x = 0$ solve the equation $45x^6 = 9x^4$?

966. What value or values for x solve the equation $x^2 - 25 = 0$?

967. What two values for x solve the equation $x^2 + 2x - 63 = 0$?

968. What two values for x solve the equation $8x^2 + 7x = 7x^2 + 8$?

969. What two values for x solve the equation $7(x^2 - x + 3) = 3(2x^2 + 2x - 7)$?

970. What two values for x solve the equation $\frac{2x+1}{15} = \frac{x^2-1}{8x}$?

Chapter 22

Solving Algebra Word Problems

- -

Algebra was invented to solve problems that can occur in the real world, so knowing how to solve equations is only half of the picture. In this chapter, you solve algebra word problems, which test your ability not only to solve equations but to set them up.

The Problems You'll Work On

Solving an algebra word problem means constructing an equation that makes sense of the jumble of information the problem contains. The problems that follow help to build the skills you need, such as:

- ✔ Declaring a variable to represent a numerical value
- ✔ Creating algebraic expressions to express values
- ✔ Constructing algebraic equations to solve word problems
- ✔ Building word problem solving skills to solve harder problems

What to Watch Out For

The key to solving an algebra word problem is clarity: Read the problem carefully so that you're clear what the problem is telling you. Then choose a variable that allows you to represent all the necessary values in the problem as expressions. Here are a few tips to help you read algebra word problems and set up algebraic equations that are correct and make sense:

- ✔ When declaring a variable, it's helpful to use a letter that makes sense. For example, use b to represent Bob's age, n to represent Nancy's savings, and so forth.

- ✔ Create an algebraic expression carefully, step by step. For example, to represent "three less than twice as much as x," first multiply x by 2, then subtract 3, resulting in the expression $2x - 3$.

- ✔ In an equation, use an equal sign to represent the word *is*. For example, represent the sentence "Three less than twice as much as x is the number 21" as $2x - 3 = 21$.

- ✔ In more difficult word problems, be sure to write down what the variable stands for at the beginning of the problem. Then record an expression to represent every other value in the problem. For example, if a problem states "Carlos has five more cookies than Ryan but three times as many as Matt," you can use c to represent Carlos. Then, be sure to record that Ryan = $c - 5$ and Matt = $3c$.

Word Problems

971–1001

971. Let *d* equal the number of dollars that you have in your savings account. If that amount doubles and then increases by an additional $1,000, how can you represent the new amount in terms of *d*?

972. Let *c* be the number of chairs in an auditorium on Saturday morning. For the day's morning event, the staff removes 20 chairs. Then for a Saturday evening event, this new number of chairs is tripled. How many chairs are now in the auditorium, in terms of *c*?

973. Let *p* equal the number of pennies that Penny has saved in her penny jar. If she removes 300 pennies from the jar today and then puts 66 pennies into the jar tomorrow, how can you represent the new number of pennies in the jar in terms of *p*?

974. Let *t* be the temperature in degrees Celsius at the base of the Eiffel Tower at 8:00 in the morning. The temperature is checked once an hour over the next few hours. The temperature first rises 5 degrees, then rises another 2 degrees, then drops 3 degrees, and finally drops another 6 degrees. What is the temperature, in terms of *t*, at the end of these changes?

975. Let *w* be the weight of a puppy at birth. Over the next few months, the weight of the puppy first triples, then increases by six pounds, and then doubles to its full adult weight. What is the final weight of the puppy in terms of *w*?

976. Let *k* equal the number of baseball cards that Kyle has in his collection. If Randy has half as many baseball cards as Kyle and Jacob has 57 more cards than Randy, how do you represent the number of cards that all three children have in terms of *k*?

977. Let *s* be the number of students at Forest Whitaker High School. Next year, 28% of the students will have graduated, and the new class of freshman will include 425 new students. Represent the number of students who will be at the school next year in terms of *s*.

978. Let *m* equal the number of miles that Millie walked on the first day of her five-day camping trip. If she increased her mileage by one mile every day of her trip, represent the total number of miles that she walked in terms of *m*.

979. Let *n* equal an odd number. What is the sum of *n* plus the next three consecutive odd numbers in terms of *n*?

980. When you multiply a number by 6 and then subtract 1, the result is 23. What is the number?

981. When you multiply a number by 3, the result is the same as when you add 8 to the number. What is the number?

982. When you multiply a number by 5, the result is the same as when you multiply the number by 3 and then add 16. What is the number?

983. When you multiply a number by 2 and then add 7, the result is the same as when you multiply the number by 3 and then add 9. What is the number?

984. When you add 6 to a number and then multiply the result by 2, the result is the same as when you multiply the number by 5 and then subtract 9. What is the number?

985. When you add 3 to a number and then multiply by 2, the result is the same as when you subtract 7 from the number and then multiply by 4. What is the number?

986. When you add 1 to a number and then divide by 3, the result is the same as when you subtract 7 from the number and then divide by 2. What is the number?

987. When you double a number, then subtract 1, and then divide the difference by 5, the result is the same as when you divide the number by 3. What is the number?

988. When you subtract 6.5 from a number and then multiply by 4, the result is the same as when you subtract 4.25 from the number. What is the number?

989. When you add 13 to a number and then divide by 4, the result is the same as when you add 11.5 to the number. What is the number?

990. When you add 1.5 to a number and then divide by 3, the result is the same as when you multiply the number by 2 and then subtract 5.5. What is the number?

991. When you square a number, subtract 12, and then divide by 4, the result is the same as when you divide the number by 2, subtract 1, and then square the result. What is the number?

992. When you multiply a number by $\frac{3}{4}$ and then subtract 2, the result is the same as when you first add 9 and then multiply the number by $\frac{2}{5}$. What is the number?

993. Lucy has $5 more than her brother Peter, and they have $27 altogether. How much money does Peter have?

994. Joannie's cellphone cost twice as much as her MP3 player and half as much as her laptop computer. If she paid $1,190 for all three together, how much did she pay for the MP3 player?

995. Cody is 8 years older than Jane and Brent is twice as many years old as Jane. If Cody is 3 years older than Brent, how old is Jane?

996. A teacher spends half of her class time going over homework problems and one-fifth of her class time reviewing for a test. Then, she uses the remaining 42 minutes for an in-class assignment. How long is the class?

997. If the sum of five consecutive integers is 165, what is the greatest of the five numbers?

998. A jar contains twice as many red marbles as blue marbles, 6 more blue marbles than yellow marbles, and one-third the number of yellow marbles as orange marbles. If it contains exactly 172 marbles, how many yellow and red marbles are in the jar?

999. A northbound train is traveling at twice the speed of a southbound train and 10 miles per hour faster than an eastbound train. If the southbound train is traveling 40 miles an hour slower than the eastbound train, how fast is the southbound train traveling?

1000. If Ken had three times as much money in his savings account and if Walter had half as much, then together they would have twice as much as they currently have. Assuming that Ken has $100 more than Walter, how much money does Ken currently have?

1001. Jessica is twice as many years old as her cousin Damar. But three years ago, she was three times older than Damar was. How old is Jessica right now?

Part II
The Answers

In this part . . .

You get the answers to all of your problems! Well, almost. Actually, you get the answers to all 1,001 problems in this book, with complete explanations to show you how to solve each correctly. If you find some of these problems to be a little over your head, or to use math skills or techniques you aren't sure about, there's more good news. I've written two more books that give you a lot more step-by-step instruction for succeeding in math:

✔ *Basic Math & Pre-Algebra For Dummies*
✔ *Basic Math & Pre-Algebra Workbook For Dummies*

And when you're finished with basic math and pre-algebra, you'll be ready for algebra. May I suggest a few additional books:

✔ *Algebra I For Dummies*
✔ *Algebra I Essentials For Dummies*
✔ *Algebra I Workbook For Dummies*
✔ *Math Word Problems For Dummies*

Visit www.dummies.com for more information.

Chapter 23

Answers

....

1. **140**

The ones digit is 6, so round up by adding 1 to the tens digit (3 + 1 = 4) and changing the 6 to 0: 136 → 140.

2. **220**

The ones digit is 4, so round down by changing the 4 to 0: 224 → 220.

3. **2,500**

The tens digit is 9, so round up by adding 1 to the hundreds digit (4 + 1 = 5) and changing both the 9 and the 2 to 0s: 2,492 → 2,500.

4. **909,100**

The tens digit is 9, so round up by adding 1 to the hundreds digit (0 + 1 = 1) and changing the 9 to 0: 909,090 → 909,100.

5. **9,000**

The hundreds digit is 0, so round down by changing every digit to the right of the thousands digit to 0: 9,099 → 9,000.

6. **235,000,000**

The hundred-thousands digit is 5, so round up by adding 1 to the millions digit (4 + 1 = 5) and changing every digit to the right of the millions digit to 0: 234,567,890 → 235,000,000.

7. **68**

Stack the numbers and add the columns from right to left.

$$\begin{array}{r} 47 \\ + 21 \\ \hline 68 \end{array}$$

8. 266

Stack the numbers and add the columns from right to left.

$$\begin{array}{r} 136 \\ 53 \\ + \ 77 \\ \hline 266 \end{array}$$

9. 2,310

Stack the numbers and add the columns from right to left.

$$\begin{array}{r} 735 \\ 246 \\ + \ 1329 \\ \hline 2310 \end{array}$$

10. 9,343

Stack the numbers and add the columns from right to left.

$$\begin{array}{r} 904 \\ 1024 \\ 6532 \\ + \ 883 \\ \hline 9343 \end{array}$$

11. 406,055

Stack the numbers and add the columns from right to left.

$$\begin{array}{r} 56702 \\ 821 \\ 5332 \\ 89 \\ + \ 343111 \\ \hline 406055 \end{array}$$

12.   **2,353,250**

Stack the numbers and add the columns from right to left.

```
  1609432
   657936
    82844
     2579
+     459
  2353250
```

13.   **35**

Stack the numbers and subtract the columns from right to left.

```
   89
 − 54
   35
```

14.   **321**

Stack the numbers and subtract the columns from right to left.

```
   373
 −  52
   321
```

15.   **172**

Stack the numbers and subtract the columns from right to left.

```
   539
 −367
   172
```

When you try to subtract the tens column, 3 is less than 6, so you must borrow 1 from the hundreds column, changing the 5 to 4. Then place this 1 in front of the 3, changing it to 13. Now subtract $13 - 6 = 7$:

```
  ⁴5̶  ¹3  9
 −  3   6  7
    1   7  2
```

16. 2,177

Stack the numbers and subtract the columns from right to left:

$$
\begin{array}{r}
2468 \\
-\ 291 \\
\hline
2177
\end{array}
$$

17. 8,333

Stack the numbers and subtract the columns from right to left:

$$
\begin{array}{r}
34825 \\
-26492 \\
\hline
8333
\end{array}
$$

18. 14,768

Stack the numbers and subtract the columns from right to left:

$$
\begin{array}{r}
71002 \\
-56234 \\
\hline
14768
\end{array}
$$

19. 1,832

Stack the first number on top of the second:

$$
\begin{array}{r}
458 \\
\times\ \ 4 \\
\hline
\end{array}
$$

Now multiply 4 by every number in 458, starting from the right. Because $4 \times 8 = 32$, a two-digit number, you write down the 2 and carry the 3 to the tens column. In the next column, you multiply $4 \times 5 = 20$ and add the 3 you carried over, giving you a total of 23. Write down the 3 and carry the 2. Multiply $4 \times 4 = 16$ and add the 2 you carried over, giving you 18:

$$
\begin{array}{r}
458 \\
\times\ \ 4 \\
\hline
1832
\end{array}
$$

20. 2,590

Stack the first number on top of the second:

$$
\begin{array}{r}
74 \\
\times 35 \\
\end{array}
$$

Now multiply 5 by every number in 74, starting from the right. Because $5 \times 4 = 20$, a two-digit number, you write down the 0 and carry the 2 to the tens column. In the next column, you multiply $5 \times 7 = 35$ and add the 2 you carried over, giving you a total of 37:

$$
\begin{array}{r}
74 \\
\times 35 \\
\hline
370
\end{array}
$$

Now multiply 3 by every number in 74, starting from the right. Because $3 \times 4 = 12$, a two-digit number, you write down the 2 and carry the 1 to the tens column. In the next column, you multiply $3 \times 7 = 21$ and add the 1 you carried over, giving you a total of 22:

$$
\begin{array}{r}
74 \\
\times 35 \\
\hline
370 \\
222
\end{array}
$$

To finish, add up the results:

$$
\begin{array}{r}
74 \\
\times \ \ 35 \\
\hline
370 \\
+222 \\
\hline
2590
\end{array}
$$

---

21.   **11,094**

Stack the first number on top of the second and multiply as shown in the answer to Question 20:

$$
\begin{array}{r}
129 \\
\times \ 86 \\
\hline
774 \\
+1032 \\
\hline
11094
\end{array}
$$

---

22.   **25,594**

Stack the first number on top of the second and multiply as shown in the answer to Question 20:

$$
\begin{array}{r}
382 \\
\times \ 67 \\
\hline
2674 \\
+2292 \\
\hline
25594
\end{array}
$$

23. **335,784**

Stack the first number on top of the second and multiply as shown in the answer to Question 20:

$$
\begin{array}{r}
9876 \\
\times\ \ 34 \\
\hline
39504 \\
+29628 \\
\hline
335784
\end{array}
$$

24. **38,062,898**

Stack the first number on top of the second and multiply as shown in the answer to Question 20:

$$
\begin{array}{r}
23834 \\
\times\ 1597 \\
\hline
166838 \\
214506 \\
119170 \\
+23834 \\
\hline
38062898
\end{array}
$$

25. **287**

Start off by writing the problem like this:

$$3\overline{)861}$$

To begin, ask how many times 3 goes into 8 — that is, what's $8 \div 3$? The answer is 2 (with a little left over), so write 2 directly above the 8. Now multiply 2×3 to get 6, place the answer directly below the 8, and draw a line beneath it:

$$
\begin{array}{r}
2 \\
3\overline{)861} \\
-6 \\
\hline
\end{array}
$$

Subtract $8 - 6$ to get 2. (*Note:* After you subtract, the result should be less than the divisor, which is 3.) Then bring down the next digit (6) to make a new number, 26:

$$
\begin{array}{r}
2 \\
3\overline{)861} \\
-6 \\
\hline
26
\end{array}
$$

These steps are one complete cycle. To complete the problem, you just need to repeat the cycle. Now ask how many times 3 goes into 26 — that is, what is $26 \div 3$? The answer is 8 (with a little left over). So write the 8 above the 6, and then multiply 8×3 to get 24. Write the answer under 26:

$$
\begin{array}{r}
28 \\
3\overline{)861} \\
-6 \\
\hline
26 \\
-24 \\
\hline
\end{array}
$$

Subtract $26 - 24$ to get 2. Then bring down the next digit (1) to make the new number, 21.

$$
\begin{array}{r}
28 \\
3\overline{)861} \\
-6 \\
\hline
26 \\
-24 \\
\hline
21 \\
\end{array}
$$

Another cycle is complete, so begin the next cycle by asking how many times 3 goes into 21 — that is, what is $21 \div 3$? The answer is 7, so write the 7 above the 1, and then multiply 7×3 to get 21. Write the answer under 21:

$$
\begin{array}{r}
287 \\
3\overline{)861} \\
-6 \\
\hline
26 \\
-24 \\
\hline
21 \\
21 \\
\end{array}
$$

Now, subtract $21 - 21 = 0$. Because you have no more numbers to bring down, you're finished, and the answer (that is, the *quotient*) is the very top number of the problem:

$$
\begin{array}{r}
287 \\
3\overline{)861} \\
-6 \\
\hline
26 \\
-24 \\
\hline
21 \\
-21 \\
\hline
0 \\
\end{array}
$$

26. 268

Use the method outlined in Question 25:

$$
\begin{array}{r}
268 \\
7\overline{)1876} \\
\underline{-14} \\
47 \\
\underline{-42} \\
56 \\
\underline{-56} \\
0
\end{array}
$$

27. 412 *r* 4

Use the method outlined in Question 25:

$$
\begin{array}{r}
412 \\
15\overline{)6184} \\
\underline{-60} \\
18 \\
\underline{-15} \\
34 \\
\underline{-30} \\
4
\end{array}
$$

The quotient is 412, and the **remainder is 4.**

28. 1,147 *r* 12

Use the method outlined in Question 25:

$$
\begin{array}{r}
1147 \\
22\overline{)25246} \\
\underline{-22} \\
32 \\
\underline{-22} \\
104 \\
\underline{-88} \\
166 \\
\underline{-154} \\
12
\end{array}
$$

The quotient is 1,147, and the **remainder is 12.**

29.   **1,132 *r* 4**

Use the method outlined in Question 25:

$$
\begin{array}{r}
1132 \\
53\overline{)60000} \\
-53 \\
\hline
70 \\
-53 \\
\hline
170 \\
-159 \\
\hline
110 \\
-106 \\
\hline
4
\end{array}
$$

The quotient is 1,132, and the remainder is 4.

30.   **1,024 *r* 1**

Use the method outlined in Question 25:

$$
\begin{array}{r}
1024 \\
256\overline{)262145} \\
-256 \\
\hline
614 \\
-512 \\
\hline
1025 \\
-1024 \\
\hline
1
\end{array}
$$

The quotient is 1,024, and the remainder is 1.

31.   **i. –3, ii. –5, iii. –1, iv. –14, v. –11**

In each case, when you subtract a smaller number minus a larger number, the result is negative. An easy way to find this result is to reverse the subtraction and then negate the result. For example, $6 - 3 = 3$, so $3 - 6 = -3$.

i. $3 - 6 = -3$

ii. $7 - 12 = -5$

iii. $14 - 15 = -1$

iv. $2 - 16 = -14$

v. $20 - 31 = -11$

32. **i. –11, ii. –10, iii. –15, iv. –17, v. –14**

In each case, when you subtract a negative number minus a positive number, you add the two numbers and the result is negative.

i. $-7 - 4 = -11$

ii. $-1 - 9 = -10$

iii. $-9 - 6 = -15$

iv. $-11 - 6 = -17$

v. $-1 - 13 = -14$

33. **i. 3, ii. –3, iii. –13, iv. 13, v. –14**

In each case, when you add a positive number plus a negative number, you subtract the smaller number from the larger, and the result has the same sign as the number that is farther from 0.

i. $-5 + 8 = 3$

ii. $-8 + 5 = -3$

iii. $-14 + 1 = -13$

iv. $-1 + 14 = 13$

v. $-20 + 6 = -14$

34. **i. –10, ii. 3, iii. –12, iv. 10, v. –20**

In each case, when you add a negative number (to either a positive or negative number), the result is the same as subtracting a positive number.

i. $-2 + (-8) = -2 - 8 = -10$

ii. $6 + (-3) = 6 - 3 = 3$

iii. $-9 + (-3) = -9 - 3 = -12$

iv. $15 + (-5) = 15 - 5 = 10$

v. $-19 + (-1) = -19 - 1 = -20$

35. **i. 6, ii. –8, iii. –7, iv. 19, v. 13**

In each case, when you subtract a negative number (from either a positive or negative number), the result is the same as adding a positive number.

i. $4 - (-2) = 4 + 2 = 6$

ii. $-9 - (-1) = -9 + 1 = -8$

iii. $-10 - (-3) = -10 + 3 = -7$

iv. $8 - (-11) = 8 + 11 = 19$

v. $-3 - (-16) = -3 + 16 = 13$

36. −64

When you add a negative number (to either a positive or negative number), the result is the same as subtracting a positive number. In this case, both values are negative, so you add the numbers and the result is negative.

$$-29 + (-35) = -29 - 35 = -64$$

37. 135

When you subtract a negative number (from either a positive or negative number), the result is the same as adding a positive number.

$$46 - (-89) = 46 + 89 = 135$$

38. −56

When you add a negative number (to either a positive or negative number), the result is the same as subtracting a positive number:

$$81 + (-137) = 81 - 137$$

Now, you're subtracting a smaller number minus a larger one, so the result is negative.

$$81 - 137 = -56$$

39. −1,154

When you subtract a negative number minus a positive number, you add the two numbers and the result is negative.

$$-212 - 942 = -1,154$$

40. −1,519

When you subtract a smaller number minus a larger number, the result is negative.

$$1,024 - 2,543 = -1,519$$

41. −10,365

When you subtract a negative number (from either a positive or negative number), the result is the same as adding a positive number.

$$-10,654 - (-289) = -10,654 + 289$$

Now, because the negative number is farther from 0 than the positive number, the result is negative.

$$-10,654 + 289 = -10,365$$

42. **See below.**

In each case, one negative in a multiplication problem results in a negative number; two negatives result in a positive number.

i. $-6 \times 9 = -54$

ii. $-8 \times (-7) = 56$

iii. $-9 \times (-7) = 63$

iv. $7 \times (-8) = -56$

v. $-9 \times (-6) = 54$

43. **–135**

When you multiply a negative number times a positive number, the result is negative.

$$-15 \times 9 = -135$$

44. **352**

When you multiply two negative numbers, the result is positive.

$$-32 \times (-11) = 352$$

45. **–1,638**

When you multiply a positive number by a negative number, the result is negative.

$$91 \times (-18) = -1,638$$

46. **210**

When you multiply an even number of negative numbers by any number of positive numbers, the result is positive. In this case, there are two negative numbers, so the product is positive.

$$-7 \times (-6) \times 5 = 210$$

47. **–400**

When you multiply an odd number of negative numbers by any number of positive numbers, the result is negative. In this case, there are three negative numbers, so the product is negative.

$$2 \times (-4) \times (-10) \times (-5) = -400$$

48. **–120**

When you multiply an odd number of negative numbers by any number of positive numbers, the result is negative. In this case, there are five negative numbers, so the product is negative.

$$-1 \times (-2) \times 3 \times (-4) \times (-5) \times (-1) = -120$$

49.　See below.

In each case, one negative in a division problem results in a negative number; two negatives result in a positive number.

i. $35 \div (-5) = -7$

ii. $-28 \div (-4) = 7$

iii. $32 \div (-4) = -8$

iv. $-48 \div -6 = 8$

v. $-36 \div 6 = -6$

50.　-22

When you divide a positive number by a negative number, the result is negative.

$$176 \div (-8) = -22$$

51.　-31

When you divide a negative number by a positive number, the result is negative.

$$-403 \div 13 = -31$$

52.　25

When you divide a negative number by another negative number, the result is positive.

$$-275 \div (-11) = 25$$

53.　62

When you divide a negative number by another negative number, the result is positive.

$$-1{,}054 \div (-17) = 62$$

54.　See below.

In each case, the absolute value of any number is its positive value; the absolute value of 0 is 0.

i. $|4-4| = |0| = 0$

ii. $|6-2| = |4| = 4$

iii. $|7-9| = |-2| = 2$

iv. $|9-1| = |8| = 8$

v. $|1-8| = |-7| = 7$

55. **61**

The absolute value of any number is its positive value:

$$|38 - 99| = |-61| = 61$$

56. **118**

The absolute value of any number is its positive value:

$$|206 - 88| = |118| = 118$$

57. **86**

The absolute value of any number is its positive value:

$$|543 - 629| = |-86| = 86$$

58. **i. 36, ii. 144, iii. 64, iv. 81, v. 1**

i. $6^2 = 6 \times 6 = 36$

ii. $12^2 = 12 \times 12 = 144$

iii. $2^6 = 2 \times 2 \times 2 \times 2 \times 2 \times 2 = 64$

iv. $3^4 = 3 \times 3 \times 3 \times 3 = 81$

v. $71^0 = 1$ (Any number [except 0] raised to the power of 0 equals 1.)

59. **676**

Express the exponent as multiplication and then evaluate.

$$26^2 = 26 \times 26 = 676$$

60. **1,728**

Express the exponent as multiplication and then evaluate the product.

$$12^3 = 12 \times 12 \times 12 = 1,728$$

61. **1,000,000**

Express the exponent as multiplication.

$$10^6 = 10 \times 10 \times 10 \times 10 \times 10 \times 10$$

Now, evaluate the multiplication.

$$10 \times 10 \times 10 \times 10 \times 10 \times 10 = 1,000,000$$

62. **3,200,000**

Express the exponent as multiplication.
$$20^5 = 20 \times 20 \times 20 \times 20 \times 20$$

Now, evaluate the multiplication.
$$20 \times 20 \times 20 \times 20 \times 20 = 3{,}200{,}000$$

63. **100,000,000**

Express the exponent as multiplication.
$$100^4 = 100 \times 100 \times 100 \times 100$$

Now, evaluate the multiplication.
$$100 \times 100 \times 100 \times 100 = 100{,}000{,}000$$

64. **1,030,301**

Express the exponent as multiplication.
$$101^3 = 101 \times 101 \times 101$$

Now, evaluate the multiplication.
$$101 \times 101 \times 101 = 1{,}030{,}301$$

65. **See below.**

i. $(-5)^2 = (-5) \times (-5) = 25$

ii. $(-4)^3 = (-4) \times (-4) \times (-4) = -64$

iii. $(-10)^5 = (-10) \times (-10) \times (-10) \times (-10) \times (-10)$
$$= -100{,}000$$

iv. $(-1)^{12} = 1$ (–1 raised to an even-numbered power always equals 1.)

v. $(-1)^{27} = -1$ (–1 raised to an odd-numbered power always equals –1.)

66. **14,641**

Express the exponent as multiplication.
$$(-11)^4 = (-11) \times (-11) \times (-11) \times (-11)$$

Now, evaluate the multiplication.
$$(-11) \times (-11) \times (-11) \times (-11) = 14{,}641$$

67. −3,375

Express the exponent as multiplication.

$$(-15)^3 = (-15) \times (-15) \times (-15)$$

Now, evaluate the multiplication.

$$(-15) \times (-15) \times (-15) = -3,375$$

68. −102,400,000

Express the exponent as multiplication.

$$(-40)^5 = (-40) \times (-40) \times (-40) \times (-40) \times (-40)$$

Now, evaluate the multiplication.

$$(-40) \times (-40) \times (-40) \times (-40) \times (-40) = -102,400,000$$

69. See below.

i. $\left(\dfrac{1}{6}\right)^2 = \dfrac{1}{6} \times \dfrac{1}{6} = \dfrac{1}{36}$

ii. $\left(\dfrac{1}{3}\right)^3 = \dfrac{1}{3} \times \dfrac{1}{3} \times \dfrac{1}{3} = \dfrac{1}{27}$

iii. $\left(\dfrac{7}{11}\right)^2 = \dfrac{7}{11} \times \dfrac{7}{11} = \dfrac{49}{121}$

iv. $\left(\dfrac{2}{5}\right)^4 = \dfrac{2}{5} \times \dfrac{2}{5} \times \dfrac{2}{5} \times \dfrac{2}{5} = \dfrac{16}{625}$

v. $\left(\dfrac{1}{10}\right)^5 = \dfrac{1}{10} \times \dfrac{1}{10} \times \dfrac{1}{10} \times \dfrac{1}{10} \times \dfrac{1}{10} = \dfrac{1}{100,000}$

70. $\dfrac{81}{484}$

First, express the exponent as fraction multiplication.

$$\left(\frac{9}{22}\right)^2 = \frac{9}{22} \times \frac{9}{22}$$

Now, solve the multiplication by multiplying the numerators (top numbers) to find the numerator of the answer and multiplying the denominators (bottom numbers) to find the denominator of the answer.

$$\frac{9}{22} \times \frac{9}{22} = \frac{9 \times 9}{22 \times 22} = \frac{81}{484}$$

71. $\dfrac{343}{27,000}$

First, express the exponent as fraction multiplication.

$$\left(\frac{7}{30}\right)^3 = \frac{7}{30} \times \frac{7}{30} \times \frac{7}{30}$$

Now, solve the multiplication by multiplying the numerators (top numbers) to find the numerator of the answer and multiplying the denominators (bottom numbers) to find the denominator of the answer.

$$\frac{7}{30} \times \frac{7}{30} \times \frac{7}{30} = \frac{7 \times 7 \times 7}{30 \times 30 \times 30} = \frac{343}{27{,}000}$$

72. $\dfrac{128}{2{,}187}$

First, express the exponent as fraction multiplication.

$$\left(\frac{2}{3}\right)^7 = \frac{2}{3} \times \frac{2}{3} \times \frac{2}{3} \times \frac{2}{3} \times \frac{2}{3} \times \frac{2}{3} \times \frac{2}{3}$$

Now, solve the multiplication by multiplying the numerators (top numbers) to find the numerator of the answer and multiplying the denominators (bottom numbers) to find the denominator of the answer.

$$= \frac{2 \times 2 \times 2 \times 2 \times 2 \times 2 \times 2}{3 \times 3 \times 3 \times 3 \times 3 \times 3 \times 3} = \frac{128}{2{,}187}$$

73. **i. 3, ii. 6, iii. 8, iv. 12, v. 17**

i. $\sqrt{9} = 3$ because $3 \times 3 = 9$.

ii. $\sqrt{36} = 6$ because $6 \times 6 = 36$.

iii. $\sqrt{64} = 8$ because $8 \times 8 = 64$.

iv. $\sqrt{144} = 12$ because $12 \times 12 = 144$.

v. $\sqrt{289} = 17$ because $17 \times 17 = 289$.

74. **i. 8, ii. 15, iii. 60, iv. 99, v. 300**

i. $2\sqrt{16} = 2 \times 4 = 8$

ii. $3\sqrt{25} = 3 \times 5 = 15$

iii. $6\sqrt{100} = 6 \times 10 = 60$

iv. $9\sqrt{121} = 9 \times 11 = 99$

v. $20\sqrt{225} = 20 \times 15 = 300$

75. $2\sqrt{2}$

Factor out the greatest possible square number from the square root:

$$\sqrt{8} = \sqrt{4}\sqrt{2}$$

Now, evaluate $\sqrt{4}$:

$$= 2\sqrt{2}$$

76. $4\sqrt{2}$

Factor out the greatest possible square number from the square root:

$$\sqrt{32} = \sqrt{16}\sqrt{2}$$

Now, evaluate $\sqrt{16}$:

$$= 4\sqrt{2}$$

77. $3\sqrt{6}$

Factor out the greatest possible square number from the square root:

$$\sqrt{54} = \sqrt{9}\sqrt{6}$$

Now, evaluate $\sqrt{9}$:

$$= 3\sqrt{6}$$

78. $4\sqrt{5}$

Factor out the greatest possible square number from the square root:

$$\sqrt{80} = \sqrt{16}\sqrt{5}$$

Now, evaluate $\sqrt{16}$:

$$= 4\sqrt{5}$$

79. $10\sqrt{3}$

Factor out the greatest possible square number from the square root:

$$\sqrt{300} = \sqrt{100}\sqrt{3}$$

Now, evaluate $\sqrt{100}$:

$$= 10\sqrt{3}$$

80. **i. 2, ii. 7, iii. 9, iv. 13, v. 20**

i. $4^{\frac{1}{2}} = \sqrt{4} = 2$

ii. $49^{\frac{1}{2}} = \sqrt{49} = 7$

iii. $81^{\frac{1}{2}} = \sqrt{81} = 9$

iv. $169^{\frac{1}{2}} = \sqrt{169} = 13$

v. $400^{\frac{1}{2}} = \sqrt{400} = 20$

81. $3\sqrt{3}$

Change the fractional exponent into a square root:

$$27^{\frac{1}{2}} = \sqrt{27}$$

Now, factor out the greatest possible square number from the square root and simplify:

$$= \sqrt{9}\sqrt{3} = 3\sqrt{3}$$

82. $2\sqrt{13}$

Change the fractional exponent into a square root:

$$52^{\frac{1}{2}} = \sqrt{52}$$

Now, factor out the greatest possible square number from the square root and simplify:

$$= \sqrt{4}\sqrt{13} = 2\sqrt{13}$$

83. $6\sqrt{2}$

Change the fractional exponent into a square root:

$$72^{\frac{1}{2}} = \sqrt{72}$$

Now, factor out the greatest possible square number from the square root and simplify:

$$= \sqrt{36}\sqrt{2} = 6\sqrt{2}$$

84. $3\sqrt{11}$

Change the fractional exponent into a square root:

$$99^{\frac{1}{2}} = \sqrt{99}$$

Now, factor out the greatest possible square number from the square root and simplify:

$$= \sqrt{9}\sqrt{11} = 3\sqrt{11}$$

85. **See below.**

To raise any number to the power of –1, place that number in the denominator (bottom number) of a fraction with a numerator (top number) of 1.

i. $3^{-1} = \frac{1}{3}$

ii. $4^{-1} = \frac{1}{4}$

iii. $10^{-1} = \frac{1}{10}$

iv. $16^{-1} = \frac{1}{16}$

v. $100^{-1} = \frac{1}{100}$

86. $\dfrac{1}{49}$

First, change the negative exponent to a positive exponent by placing the number in the denominator of a fraction:

$$7^{-2} = \frac{1}{7^2}$$

Now, evaluate the exponent:

$$= \frac{1}{49}$$

87. $\dfrac{1}{64}$

First, change the negative exponent to a positive exponent by placing the number in the denominator of a fraction:

$$2^{-6} = \frac{1}{2^6}$$

Now, evaluate the exponent:

$$= \frac{1}{64}$$

88. $\dfrac{1}{625}$

First, change the negative exponent to a positive exponent by placing the number in the denominator of a fraction:

$$5^{-4} = \frac{1}{5^4}$$

Now, evaluate the exponent:

$$= \frac{1}{625}$$

89. $\dfrac{1}{169}$

First, change the negative exponent to a positive exponent by placing the number in the denominator of a fraction:

$$13^{-2} = \frac{1}{13^2}$$

Now, evaluate the exponent:

$$= \frac{1}{169}$$

90. $\dfrac{1}{1{,}000{,}000}$

First, change the negative exponent to a positive exponent by placing the number in the denominator of a fraction:

$$10^{-6} = \frac{1}{10^6}$$

Now, evaluate the exponent:

$$= \frac{1}{1{,}000{,}000}$$

91. **14**

All operations are addition and subtraction, so evaluate from left to right:

$$8 + 9 - 3$$
$$= 17 - 3$$
$$= 14$$

92. **–16**

All operations are addition and subtraction, so evaluate from left to right:

$$-5 - 10 + 3 - 4$$
$$= -15 + 3 - 4$$
$$= -12 - 4$$
$$= -16$$

93. **3**

All operations are multiplication and division, so evaluate from left to right:

$$4 \times 6 \div 8$$
$$= 24 \div 8$$
$$= 3$$

94. **8**

All operations are multiplication and division, so evaluate from left to right:

$$28 \div 7 \times 4 \div 2$$
$$= 4 \times 4 \div 2$$
$$= 16 \div 2$$
$$= 8$$

95. 30

All operations are multiplication and division, so evaluate from left to right:

$$-35 \div 7 \times (-6)$$
$$= -5 \times (-6)$$
$$= 30$$

96. 16

All operations are multiplication and division, so evaluate from left to right:

$$72 \div (-9) \times (-4) \div 2$$
$$= -8 \times (-4) \div 2$$
$$= 32 \div 2$$
$$= 16$$

97. 9

Operations include both multiplication/division and addition/subtraction. Begin by evaluating all multiplication/division from left to right:

$$56 \div 7 + 1 = 8 + 1$$

Next, evaluate all addition/subtraction:

$$= 9$$

98. −1

Operations include both multiplication/division and addition/subtraction. Begin by evaluating all multiplication/division from left to right:

$$15 - 8 \times 2 = 15 - 16$$

Next, evaluate all addition/subtraction:

$$= -1$$

99. 16

Operations include both multiplication/division and addition/subtraction. Begin by evaluating all multiplication/division from left to right:

$$12 + 10 \div 2 - 1 = 12 + 5 - 1$$

Next, evaluate all addition/subtraction from left to right:

$$= 17 - 1$$
$$= 16$$

100.   26

Operations include both multiplication/division and addition/subtraction. Begin by evaluating all multiplication/division from left to right:

$$18 + 36 \div 9 \times 2$$
$$= 18 + 4 \times 2$$
$$= 18 + 8$$

Next, evaluate all addition/subtraction:

$$= 26$$

101.   –41

Operations include both multiplication/division and addition/subtraction. Begin by evaluating all multiplication/division from left to right:

$$75 \div (-5) \times 3 + 4$$
$$= -15 \times 3 + 4$$
$$= -45 + 4$$

Next, evaluate all addition/subtraction:

$$= -41$$

102.   –54

Operations include both multiplication/division and addition/subtraction. Begin by evaluating all multiplication/division from left to right:

$$-6 \times 7 + (-36) \div 3$$
$$= -42 + (-36) \div 3$$
$$= -42 + (-12)$$

Next, evaluate all addition/subtraction:

$$-42 - 12 = -54$$

103.   400

Operations include both exponents and multiplication/division. Begin by evaluating all exponents from left to right:

$$4 \times 10^2$$
$$= 4 \times 100$$

Next, evaluate all multiplication/division:

$$= 400$$

104. 140

Operations include both exponents and multiplication/division. Begin by evaluating all exponents:

$$56 \div 2^3 \times 20$$
$$= 56 \div 8 \times 20$$

Next, evaluate all multiplication/division from left to right:

$$= 7 \times 20$$
$$= 140$$

105. 22

Operations include both exponents and addition/subtraction. Begin by evaluating all exponents:

$$1 + 5^2 - 4$$
$$= 1 + 25 - 4$$

Next, evaluate all addition/subtraction from left to right:

$$= 26 - 4$$
$$= 22$$

106. 25

Operations include both exponents and addition/subtraction. Begin by evaluating all exponents from left to right:

$$3^3 + 2^3 - 10$$
$$= 27 + 2^3 - 10$$
$$= 27 + 8 - 10$$

Next, evaluate all addition/subtraction from left to right:

$$= 35 - 10$$
$$= 25$$

107. –23

Operations include both exponents and addition/subtraction. Begin by evaluating all exponents from left to right:

$$-2^5 + 3^2$$
$$= -32 + 3^2$$
$$= -32 + 9$$

Next, evaluate all addition/subtraction:

$$= -23$$

108. 46

Operations include exponents, multiplication/division, and addition/subtraction. Begin by evaluating all exponents from left to right:

$$7^2 - 6^0 \times 3$$
$$= 49 - 6^0 \times 3$$
$$= 49 - 1 \times 3$$

Next, evaluate all multiplication/division:

$$= 49 - 3$$

Finally, evaluate all addition/subtraction:

$$= 46$$

109. 0

Operations include exponents, multiplication/division, and addition/subtraction. Begin by evaluating all exponents from left to right:

$$10^5 \div 10^4 - 10$$
$$= 100,000 \div 10^4 - 10$$
$$= 100,000 \div 10,000 - 10$$

Next, evaluate all multiplication/division:

$$= 10 - 10$$

Finally, evaluate all addition/subtraction:

$$= 0$$

110. 500

Operations include exponents, multiplication/division, and addition/subtraction. Begin by evaluating all exponents from left to right:

$$-20 \times 25 + 2^3 \times 5^3$$
$$= -20 \times 25 + 8 \times 5^3$$
$$= -20 \times 25 + 8 \times 125$$

Next, evaluate all multiplication/division from left to right:

$$= -500 + 8 \times 125$$
$$= -500 + 1000$$

Finally, evaluate all addition/subtraction:

$$= 500$$

111. 120

Operations include exponents, multiplication/division, and addition/subtraction. Begin by evaluating all exponents from left to right:

$$(-8)^2 \div 2^3 \times 40 + (-200)$$
$$= 64 \div 2^3 \times 40 + (-200)$$
$$= 64 \div 8 \times 40 + (-200)$$

Next, evaluate all multiplication/division from left to right:

$$= 8 \times 40 + (-200)$$
$$= 320 + (-200)$$

Finally, evaluate all addition/subtraction:

$$= 120$$

112. 5

Operations include exponents, multiplication/division, and addition/subtraction. Begin by evaluating all exponents from left to right:

$$-1^3 \times (-2) + 9^2 \div 3^3$$
$$= -1 \times (-2) + 9^2 \div 3^3$$
$$= -1 \times (-2) + 81 \div 3^3$$
$$= -1 \times (-2) + 81 \div 27$$

Next, evaluate all multiplication/division from left to right:

$$= 2 + 81 \div 27$$
$$= 2 + 3$$

Finally, evaluate all addition/subtraction:

$$= 5$$

113. 147

Begin by evaluating the expression inside the parentheses:

$$7^2 \times (6 - 3)$$
$$= 7^2 \times 3$$

Remaining operations include exponents and multiplication/division. Begin by evaluating all exponents:

$$= 49 \times 3$$

Next, evaluate all multiplication/division:

$$= 147$$

114.   –75

Begin by evaluating the expression inside the parentheses using the proper order of operations:

$$5 \times (3 - 9 \times 2)$$
$$= 5 \times (3 - 18)$$
$$= 5 \times (-15)$$

Next, evaluate the multiplication:

$$= -75$$

115.   –1

Begin by evaluating the expressions inside each set of parentheses, starting with the first, using the proper order of operations in each case:

$$(-9 \div 3) \div ((-6)^2 \div 12)$$
$$= -3 \div (-6^2 \div 12)$$
$$= -3 \div (36 \div 12)$$
$$= -3 \div 3$$

Next, evaluate the remaining division:

$$= -1$$

116.   28

Begin by evaluating the expression inside the first set of parentheses, using the proper order of operations:

$$(5 \times 3 - 1) \times (50 \div 5^2)$$
$$= (15 - 1) \times (50 \div 5^2)$$
$$= 14 \times (50 \div 5^2)$$

Next, evaluate the expression inside the second set of parentheses, using the proper order of operations:

$$= 14 \times (50 \div 25)$$
$$= 14 \times 2$$

Finally, evaluate the remaining multiplication:

$$= 28$$

117. 2

Begin by evaluating the expressions inside each set of parentheses, starting with the first, using the proper order of operations in each case:

$$(11-3)^2 \div \left(6^2 - 4\right)$$
$$= 8^2 \div \left(6^2 - 4\right)$$
$$= 8^2 \div \left(36 - 4\right)$$
$$= 8^2 \div 32$$

Next, evaluate the exponent and then the division:

$$= 64 \div 32 = 2$$

118. −123

Begin by evaluating the expression inside the brackets:

$$\left[12 \div (-4)\right] \times \left(10 \times 2^2 + 1\right)$$
$$= -3 \times \left(10 \times 2^2 + 1\right)$$

Next, evaluate the expression inside the parentheses, using the proper order of operations:

$$= -3 \times \left(10 \times 4 + 1\right)$$
$$= -3 \times \left(40 + 1\right)$$
$$= -3 \times 41$$

Finally, evaluate the remaining multiplication:

$$= -123$$

119. 20

Begin by evaluating the expression inside the first set of parentheses, using the proper order of operations:

$$\left(-1^3 - 5\right)^2 - \left(6 - 4 \div 2\right)^2$$
$$= \left(-1 - 5\right)^2 - \left(6 - 4 \div 2\right)^2$$
$$= \left(-6\right)^2 - \left(6 - 4 \div 2\right)^2$$

Next, evaluate the expression inside the second set of parentheses:

$$= \left(-6\right)^2 - \left(6 - 2\right)^2$$
$$= \left(-6\right)^2 - \left(4\right)^2$$

Next, evaluate the exponents:

$$= 36 - \left(4\right)^2$$
$$= 36 - 16$$

Finally, evaluate the subtraction:

$$= 20$$

120. 13

Begin by evaluating the expression in the innermost set of parentheses:

$$\left[(5-2)\times 4\right]+1$$
$$=\left[3\times 4\right]+1$$

Next, evaluate the expression inside the remaining set of parentheses and then evaluate the addition:

$$=12+1=13$$

121. 86

Begin by evaluating the expression in the innermost set of parentheses:

$$50-\left[(-6+2)\times 3^2\right]$$
$$=50-\left[-4\times 3^2\right]$$

Next, evaluate the expression inside the remaining set of parentheses, using the proper order of operations:

$$=50-\left[-4\times 9\right]$$
$$=50-\left[-36\right]$$

Finally, evaluate the subtraction:

$$=50+36=86$$

122. 300

Begin by evaluating the expression in the innermost set of parentheses:

$$3\times\left[4\times(-3+8)^2\right]$$
$$=3\times\left[4\times(-5)^2\right]$$

Next, evaluate the expression inside the remaining set of parentheses:

$$=3\times\left[4\times 25\right]$$
$$=3\times 100$$

Finally, evaluate the multiplication:

$$=300$$

123. −2

Begin by evaluating the expressions in the two innermost sets of parentheses, starting with the first:

$$-24\div\left[(-2-10)\times\left(7-2^3\right)\right]$$
$$=-24\div\left[-12\times\left(7-2^3\right)\right]$$

Next, evaluate the expression inside the remaining innermost set of parentheses, using the proper order of operations:

$$= -24 \div \left[-12 \times (7-8) \right]$$
$$= -24 \div \left[-12 \times (-1) \right]$$

Now, evaluate the expression inside the remaining set of parentheses:

$$= -24 \div 12$$

Finally, evaluate the division:

$$= -2$$

124. 6

Begin by evaluating the expression in the innermost set of parentheses and work your way outward:

$$4 + \left\{ \left[(5-1) \times 7 \right] \div 14 \right\}$$
$$= 4 + \left\{ \left[4 \times 7 \right] \div 14 \right\}$$
$$= 4 + \left\{ 28 \div 14 \right\}$$
$$= 4 + 2$$
$$= 6$$

125. 2

Begin by evaluating the square root and then evaluate the division:

$$\sqrt{64} \div 4$$
$$= 8 \div 4$$
$$= 2$$

126. 4

Begin by evaluating the two square roots:

$$\sqrt{100} - \sqrt{36}$$
$$= 10 - \sqrt{36}$$
$$= 10 - 6$$

Then, evaluate the subtraction:

$$= 4$$

127. 0

Begin by evaluating the square root:

$$-1 + \sqrt{81} \div 9$$
$$= -1 + 9 \div 9$$

Next, evaluate the division and finally the addition:

$$= -1 + 1 = 0$$

128. 7

Begin by evaluating the expression inside the square root:

$$\sqrt{4 \times 9} \div 2 + 4$$
$$= \sqrt{36} \div 2 + 4$$

Next, evaluate the square root:

$$= 6 \div 2 + 4$$

Next, evaluate the division and finally the addition:

$$= 3 + 4 = 7$$

129. −13

Begin by evaluating the expression inside the square root, evaluating the division first and then the addition:

$$-8 - \sqrt{24 + 3 \div 3}$$
$$= -8 - \sqrt{24 + 1}$$
$$= -8 - \sqrt{25}$$

Next, evaluate the square root and then the subtraction:

$$= -8 - 5 = -13$$

130. 7

Begin by evaluating the expression inside the square root, starting inside the parentheses:

$$\sqrt{-20 \div (-7 + 2)} + 5$$
$$= \sqrt{-20 \div (-5)} + 5$$
$$= \sqrt{4} + 5$$

Next, evaluate the square root and then the addition:

$$= 2 + 5 = 7$$

131. 1

Begin by evaluating the expression inside the square root, starting with the exponent and then the multiplication:

$$\sqrt{79 - 5 \times 2^4 + 2}$$
$$= \sqrt{79 - 5 \times 16 + 2}$$
$$= \sqrt{79 - 80 + 2}$$

Next, evaluate the addition/subtraction from left to right:

$$= \sqrt{-1+2}$$

$$= \sqrt{1}$$

Finally, evaluate the square root:

$$=1$$

132. **19**

Begin by evaluating the expression inside the first square root, starting with the exponents and then the multiplication:

$$\sqrt{5^2 \times 3^2} + \sqrt{5^2 - 3^2}$$

$$= \sqrt{25 \times 9} + \sqrt{5^2 - 3^2}$$

$$= \sqrt{225} + \sqrt{5^2 - 3^2}$$

Next, evaluate the expression inside the second square root, starting with the exponents and then the subtraction:

$$= \sqrt{225} + \sqrt{25 - 9}$$

$$= \sqrt{225} + \sqrt{16}$$

Finally, evaluate the two square roots and then the addition:

$$= 15 + 4 = 19$$

133. **100**

Begin by evaluating the square root, using the proper order of operations:

$$\left[\sqrt{(13+5) \times 2} - 4^2 \right]^2$$

$$= \left[\sqrt{18 \times 2} - 4^2 \right]^2$$

$$= \left[\sqrt{36} - 4^2 \right]^2$$

$$= \left[6 - 4^2 \right]^2$$

Next, evaluate the expression inside the parentheses, starting with the exponent and then the subtraction:

$$= \left[6 - 16 \right]^2$$

$$= \left[-10 \right]^2$$

Finally, evaluate the exponent:

$$= 100$$

134. −27

Begin by evaluating the innermost square root, using the proper order of operations:

$$\left(\sqrt{4+\sqrt{4^2\times2^4\times2}}-3^2\right)^3$$
$$=\left(\sqrt{4+\sqrt{16\times16\times2}}-3^2\right)^3$$
$$=\left(\sqrt{4+\sqrt{256\times2}}-3^2\right)^3$$
$$=\left(\sqrt{4+16\times2}-3^2\right)^3$$

Now, evaluate the remaining square root:

$$=\left(\sqrt{4+32}-3^2\right)^3$$
$$=\left(\sqrt{36}-3^2\right)^3$$
$$=\left(6-3^2\right)^3$$

Next, evaluate the value in the parentheses:

$$=(6-9)^3$$
$$=(-3)^3$$

Finally, evaluate the exponent

$$=-27$$

135. 3

Begin by evaluating the expressions in the numerator (top number) and denominator (bottom number) of the fraction:

$$\frac{8-2}{16\div8}=\frac{6}{2}$$

Next, evaluate the fraction by dividing the numerator by the denominator:

$$=6\div2=3$$

136. −4

Begin by evaluating the expressions in the numerator (top number) of the fraction:

$$\frac{4\times\sqrt{25}}{-7-(-2)}$$
$$=\frac{4\times5}{-7-(-2)}$$
$$=\frac{20}{-7-(-2)}$$

Now, evaluate the denominator (bottom number):

$$= \frac{20}{-7+2}$$

$$= \frac{20}{-5}$$

Next, evaluate the fraction by dividing the numerator by the denominator:

$$= 20 \div -5 = -4$$

137.   1

Begin by evaluating all of the exponents in the numerator (top number) of the fraction, and then evaluating the addition and subtraction from left to right:

$$\frac{2^5 - 2^4 + 2^3}{2^4 + 2^3}$$

$$= \frac{32 - 16 + 8}{2^4 + 2^3}$$

$$= \frac{24}{2^4 + 2^3}$$

Now, evaluate the denominator (bottom number) in the same way:

$$= \frac{24}{16+8}$$

$$= \frac{24}{24}$$

Next, evaluate the fraction by dividing the numerator by the denominator:

$$= 24 \div 24 = 1$$

138.   −2

Begin by evaluating the expressions in the numerator (top number):

$$\frac{3^3 + 17}{-6 + (-8 \times 2)}$$

$$= \frac{27 + 17}{-6 + (-8 \times 2)}$$

$$= \frac{44}{-6 + (-8 \times 2)}$$

Next, evaluate the denominator (bottom number) of the fraction:

$$= \frac{44}{-6 + (-8 \times 2)}$$

$$= \frac{44}{-6 + (-16)}$$

$$= \frac{44}{-22}$$

Now, evaluate the fraction by dividing the numerator by the denominator:

$$= 44 \div (-22) = -2$$

139.   2

Begin by evaluating the expressions in the numerator (top number):

$$\frac{\sqrt{2^5-(-4)}}{\left[12-(1-7)\right]\div 6}$$

$$=\frac{\sqrt{32-(-4)}}{\left[12-(1-7)\right]\div 6}$$

$$=\frac{\sqrt{36}}{\left[12-(1-7)\right]\div 6}$$

$$=\frac{6}{\left[12-(1-7)\right]\div 6}$$

Next, evaluate the denominator (bottom number):

$$=\frac{6}{\left[12-(-6)\right]\div 6}$$

$$=\frac{6}{18\div 6}$$

$$=\frac{6}{3}$$

Evaluate the fraction by dividing the numerator by the denominator:

$$=6\div 3=2$$

140.   2

Begin by evaluating the expression inside the inner parentheses in the numerator (top number):

$$\sqrt{\frac{\left[22\div\left(7+2^2\right)\right]+\left(7^2-15\right)}{\sqrt{4^3+2^4-\left(-1^3\right)}}}$$

$$=\sqrt{\frac{\left[22\div(7+4)\right]+\left(7^2-15\right)}{\sqrt{4^3+2^4-\left(-1^3\right)}}}$$

$$=\sqrt{\frac{\left[\left[22\div 11\right]+\left(7^2-15\right)\right.}{\sqrt{4^3+2^4-\left(-1^3\right)}}}$$

$$=\sqrt{\frac{2+\left(7^2-15\right)}{\sqrt{4^3+2^4-\left(-1^3\right)}}}$$

Continue by evaluating the numerator:

$$= \sqrt{\frac{2+(49-15)}{\sqrt{\sqrt{4^3+2^4-(-1^3)}}}}$$

$$= \sqrt{\frac{2+34}{\sqrt{\sqrt{4^3+2^4-(-1^3)}}}}$$

$$= \sqrt{\frac{36}{\sqrt{\sqrt{4^3+2^4-(-1^3)}}}}$$

Next, evaluate the denominator (bottom number):

$$= \sqrt{\frac{36}{\sqrt{\sqrt{64+16-(-1)}}}}$$

$$= \sqrt{\frac{36}{\sqrt{\sqrt{64+16+1}}}}$$

$$= \sqrt{\frac{36}{\sqrt{\sqrt{81}}}}$$

$$= \sqrt{\frac{36}{9}}$$

Finally, evaluate the fraction and the remaining square root:

$$= \sqrt{4} = 2$$

141. −6

Begin by evaluating the expression inside the absolute value:

$$|-8+2\times(-5)| \div (-3)$$
$$= |-8-10| \div (-3)$$
$$= |-18| \div (-3)$$

Now, evaluate the absolute value and division:

$$= 18 \div (-3) = -6$$

142. −20

Begin by evaluating the expression inside the absolute value, starting inside the parentheses:

$$|(7-11) \div 2| \times (3-13)$$
$$= |-4 \div 2| \times (3-13)$$
$$= |-2| \times (3-13)$$
$$= 2 \times (3-13)$$

Now, evaluate the expression inside the remaining parentheses and then the multiplication:

$$= 2 \times (-10) = -20$$

143. **56**

Begin by evaluating the expression inside the first absolute value:

$$|4-9| \times (17-5) - |8 \div (-2)|$$
$$= |-5| \times (17-5) - |8 \div (-2)|$$
$$= 5 \times (17-5) - |8 \div (-2)|$$

Next, evaluate the expression inside the parentheses:

$$= 5 \times 12 - |8 \div (-2)|$$

Now, evaluate the expression inside the remaining absolute value:

$$= 5 \times 12 - |-4|$$
$$= 5 \times 12 - 4$$

To finish, evaluate the multiplication and then the subtraction:

$$= 60 - 4 = 56$$

144. **–1**

Begin by evaluating the numerator (top number) of the fraction. Start with the first absolute value inside the square root:

$$\frac{\sqrt{|44-85| + |(5-70) \div 13| - (-3)}}{[7 \div (10-3)] + (-8)}$$

$$= \frac{\sqrt{|-41| + |(5-70) \div 13| - (-3)}}{[7 \div (10-3)] + (-8)}$$

$$= \frac{\sqrt{41 + |(5-70) \div 13| - (-3)}}{[7 \div (10-3)] + (-8)}$$

Continue by evaluating the remaining absolute value inside the square root:

$$= \frac{\sqrt{41 + |-65 \div 13| - (-3)}}{[7 \div (10-3)] + (-8)}$$

$$= \frac{\sqrt{41 + |-5| - (-3)}}{[7 \div (10-3)] + (-8)}$$

$$= \frac{\sqrt{41 + 5 - (-3)}}{[7 \div (10-3)] + (-8)}$$

Now, evaluate the square root:

$$= \frac{\sqrt{46 - (-3)}}{[7 \div (10-3)] + (-8)}$$

$$= \frac{\sqrt{49}}{[7 \div (10-3)] + (-8)}$$

$$= \frac{7}{[7 \div (10-3)] + (-8)}$$

Now, evaluate the denominator (bottom number) of the fraction, starting with the innermost parentheses:

$$= \frac{7}{[7 \div 7] + (-8)}$$

$$= \frac{7}{1 + (-8)}$$

$$= \frac{7}{-7}$$

To finish, evaluate the fraction as division:

$$7 \div (-7) = -1$$

145. 290 minutes

The three movies are 80 minutes, 95 minutes, and 115 minutes, so calculate as follows:

$$80 + 95 + 115 = 290$$

146. 1,454 feet

The Burj Khalifa is 2,717 feet, and it is 1,263 feet taller than the Empire State Building, so subtract as follows:

$$2,717 \text{ feet} - 1,263 \text{ feet} = 1,454 \text{ feet}$$

147. 10

Five dozen eggs equals 60 eggs altogether (because $12 \times 5 = 60$). These 60 eggs are divided evenly among six children, so each child receives $60 \div 6 = 10$ eggs.

148. $90

The first week, Arturo worked for 40 hours at $12 per hour, so he earned $12 \times 40 = \$480$. The second week he worked for 30 hours at $13 per hour, so he earned $13 \times 30 = \$390$. To figure out how much more money he received for the first week of work, calculate as follows:

$$\$480 - \$390 = \$90$$

Therefore, he earned $90 more the first week than the second.

149. 180 people

The restaurant has 5 tables that seat 8 people, so these tables hold $5 \times 8 = 40$ people. It has 16 tables that seat 6 people, so these tables hold $16 \times 6 = 96$ people. And it has 11 tables that seat 4 people, so these tables hold $11 \times 4 = 44$ people. The total capacity of all the tables is the sum of their totals. Calculate as follows:

$$40 + 96 + 44 = 180$$

Therefore, the tables hold 180 people altogether.

150. **320 pounds**

A pint of water weighs 1 pound, and a gallon contains 8 pints, so a gallon of water weighs 8 pounds. Thus, 40 gallons of water weigh $40 \times 8 = 320$ pounds.

151. **$23**

The sweater sold for $86, with half off, so Antonia bought it for $86 ÷ 2 = $43. She used a $20 gift card to pay, so she spent $43 – $20 = $23 of her own money.

152. **$4.50**

Almonte and Karan both bought six notebooks. But Karan bought two large notebooks and Almonte bought five of them. Thus, Almonte bought three more large notebooks than Karan, so he spent $1.50 × 3 = $4.50 more than Karan.

153. **8 months**

Each sale yields a $35 return on a $7,000,000 investment, so 200,000 sales are needed to break even (because 7,000,000 ÷ 35 = 200,000). And 25,000 are sold each month, so it will take 8 months to pay off the investment (200,000 ÷ 25,000 = 8).

154. **$3**

Jessica wants to buy 40 pens, so she either has to buy 5 packs of 8 pens (because $8 \times 5 = 40$) or 4 packs of 10 pens (because $10 \times 4 = 40$). If she buys 5 packs of 8 pens at $7 per pack, she pays $7 \times 5 = $35. However, if she buys 4 packs of 10 at $8 per pack, she pays $8 \times 4 = $32. Thus, Jessica saves $35 – $32 = $3.

155. **58**

Jim bought one 10-ounce box and three 16-ounce boxes, so calculate as follows:

$$(1 \times 10) + (3 \times 16) = 10 + 48 = 58$$

156. **32**

Mina walked 3 miles on each of 4 days, and 5 miles on each of the other 4 days, so she walked a total of:

$$(3 \times 4) + (5 \times 4) = 12 + 20 = 32 \text{ miles.}$$

157. **70**

The bike-a-thon is a total of 250 miles. The first day accounts for 100 miles and the second day amounts to 100 – 20 = 80 miles, so the distance of the third day is:

$$250 - 100 - 80 = 70$$

158. $63

Six T-shirts sell for $42, so each T-shirt sells for $42 ÷ 6 = $7. Calculate the cost of nine T-shirts as follows:

$7 × 9 = $63

159. 135

Kenny did 25 pushups, Sal did 25 × 2 = 50 pushups, and Natalie did 50 + 10 = 60 pushups. Calculate the total combined pushups as follows:

25 + 50 + 60 = 135

160. 10 cents

To find the price of 1 candy bar when you buy a package of 2, divide 90 by 2:

90 ÷ 2 = 45

To find the price of 1 candy bar when you buy a package of 3 for $1.05, divide 105 by 3:

105 ÷ 3 = 35

Therefore, you can save 45 – 35 = 10 cents per candy bar by buying a package of 3.

161. 49

The two square numbers are less than 130, so neither is greater than 121. Also, the larger of the two square numbers must be at least 64, because if not, then the two numbers couldn't add up to 130. Begin by subtracting square numbers from 130 to find pairs of square numbers that work:

130 – 121 = 9

130 – 100 = 30

130 – 81 = 49

130 – 64 = 66

So the two possible pairs that add up to 130 are:

9 + 121 = 130

49 + 81 = 130

But 121 – 9 = 112, so this is not right. However, 81 – 49 = 32, so the right answer is 49.

162. 96 minutes

Donna read 60 pages in 20 minutes, so she took one minute to read 3 pages (because 60 ÷ 20 = 3). Thus, calculate as follows to find out how long she took to read 288 pages:

288 ÷ 3 = 96

Thus, she took 96 minutes.

163. 500 boxes

Kendra sold 50 boxes of cookies in 20 days, so at this rate she would have sold 100 boxes in 40 days. Alicia sold twice as many boxes in half as many days, so she sold 100 boxes in 10 days. Thus, at the same rate, she would have sold 400 boxes in 40 days. Therefore, together they would have sold 100 + 400 = 500 boxes of cookies in 40 days.

164. 5

The group of 70 contains 3 girls for every 4 boys, so it contains 30 girls and 40 boys. Of these, 6 girls pair up with 6 boys, leaving 24 girls and 34 boys. These children form 12 pairs of girls (24 ÷ 2 = 12) and 17 pairs of boys (34 ÷ 2 = 17). Therefore, there are 5 more boy-boy pairs than girl-girl pairs.

165. 21

The book and the newspaper cost $11.00 together, and the book costs $10.00 more than the newspaper. Thus, the book costs $10.50 and the newspaper costs $.50. So, you could buy 21 newspapers for the same price as the book (because 1,050 ÷ 50 = 21).

166. $263,000

Yianni's payment is $1,800 over 30 years, with 12 payments per year, so calculate as follows:

$$1,800 \times 30 \times 12 = \$648,000$$

This price is over and above the principal of $385,000, so calculate total interest as follows:

$$\$648,000 - \$385,000 = \$263,000$$

Therefore, Yianni will have paid $263,000 in interest.

167. 450 mph

The plane traveling west-to-east from San Diego to New York moves at a forward speed of 540 miles per hour. Thus, you can calculate its time in hours by dividing the total distance by the speed it travels as follows:

$$2,700 \div 540 = 5$$

So, this plane travels for five hours. When traveling from east-to-west, the flight takes one hour longer, so it takes six hours. Thus, you can calculate forward speed in miles per hour by dividing the total distance by the time it takes as follows:

$$2,700 \div 6 = 450$$

Therefore, under the same conditions, a plane traveling from New York to San Diego moves at a forward speed of 450 miles per hour.

168. Won $40

Arlo lost $65, then won $120, then lost $45, and then won $30, so calculate as follows:

$$-\$65 + \$120 - \$45 + 30 = \$40$$

Therefore, Arlo won $40.

169. $400

Clarissa bought the diamond for $1,000 and sold it for $1,100, so she made a profit of $100 in this first transaction. Later, she bought the same diamond again for $900 and sold it for $1,200, so she made a further profit of $300. Therefore, she made a total profit of $400.

170. 96 minutes

Angela can make 4 sandwiches in 3 minutes, so at this rate she can make 16 sandwiches in 12 minutes. Basil can make 3 sandwiches in 4 minutes, so at this rate he can make 9 sandwiches in 12 minutes. Thus, working together, they can make 25 sandwiches in 12 minutes. If they do this 8 times in a row, they can make 200 sandwiches (because $25 \times 8 = 200$), and this will take 96 minutes (because $12 \times 8 = 96$).

171. 9

Each of the 16 children has at least 2 siblings, so this accounts for $16 \times 2 = 32$ siblings. There are 41 siblings in total, so $41 - 32 = 9$ siblings who are not accounted for. Thus, 9 of the children in Ms. Morrow's class have an extra sibling, so these 9 children have exactly 3 siblings.

172. 5,050

Begin by pairing the first and last numbers, the second and second-to-last numbers, the third and third-to-last numbers, and so forth:

$$1 + 100 = 101$$
$$2 + 99 = 101$$
$$3 + 98 = 101$$
$$\ldots$$

All of these pairings total 101. Furthermore, notice that you can continue these pairings until you reach the middle two numbers:

$$\ldots$$
$$48 + 53 = 101$$
$$49 + 52 = 101$$
$$50 + 51 = 101$$

As you can see, the result is 50 pairs of numbers, each of which adds up to 101. So you can find the total value of all the pairings as follows:

$$101 \times 50 = 5,050$$

173. $6,850

Louise's quota was $1,200 per day, so it was $1,200 × 5 = $6,000 for the five days. She exceeded this amount by $450 on Monday and $650 on Tuesday, and missed it by $250 on Friday, so calculate her weekly total as follows:

$$\$6{,}000 + \$450 + \$650 - \$250 = \$6{,}850$$

174. 30 seconds

Assuming that he could continue swimming at a constant rate of 3 miles per hour, Jordy could have gone 120 lengths of the pool in 1 hour (because 3 × 40 = 120). Thus, at that rate, he could swim 2 lengths per minute (because 120 ÷ 60 = 2), or 1 length in 30 seconds.

175. 45

To begin, pick a person from the group and call him Person #1. The group has 10 people, so Person #1 can shake hands with 9 different people (People #2 through #10). This accounts for all handshakes that involve Person #1, so set him aside.

Now, consider that Person #2 can shake hands with 8 different people (People #3 through #10) before being set aside. And Person #3 can shake hands with 7 different people, and so on. This goes on until Person #9 shakes hands with Person #10. At that point, all pairings are complete.

So, calculate this as follows:

$$9 + 8 + 7 + 6 + 5 + 4 + 3 + 2 + 1 = 45$$

176. 6 pounds

The first measurement that Marion made was as follows:

3 red bricks + 1 white brick = 23 pounds

After she removed one red brick and added two white bricks, the second measurement was 4 pounds greater:

2 red bricks + 3 white bricks = 27 pounds

Thus, each time she removes one red brick and adds two white bricks, you can expect the weight to increase by 4 pounds:

1 red brick + 5 white bricks = 31 pounds

7 white bricks = 35 pounds

Thus, because 35 ÷ 7 = 5:

1 white brick = 5 pounds

5 white bricks = 25 pounds

Therefore:

1 red brick + 5 white bricks = 31 pounds

1 red brick + 25 pounds = 31 pounds

1 red brick = 6 pounds

177. $60.36

Calculate the four amounts by multiplying the number of coins times the value of each type of coin as follows:

$$891 \times 0.1 = 8.91$$
$$342 \times 0.5 = 17.10$$
$$176 \times 0.10 = 17.60$$
$$67 \times 0.25 = 16.75$$

Now, add up the results:

$$8.91 + 17.10 + 17.60 + 16.75 = 60.36$$

Therefore, the jar contained a total of $60.36.

178. 55 mph

Joel drove for a total of 2 hours + 1 hour + 2 hours + 3 hours = 8 hours. He took a total of 15 minutes + 45 minutes = 60 minutes = 1 hour for breaks, so his total time for the trip was 8 hours + 1 hour = 9 hours.

He drove at 70 mph for 2 hours, 60 mph for 1 hour, 35 mph for 2 hours, and 75 mph for 3 hours, so calculate his total distance as follows:

$$(70 \times 2) + (60 \times 1) + (35 \times 2) + (75 \times 3)$$
$$= 140 + 60 + 70 + 225$$
$$= 495$$

Thus, he drove 495 miles in 9 hours. To find the average speed, divide 495 by 9:

$$495 \div 9 = 55$$

Therefore, his average speed for the whole trip, including breaks, was 55 mph.

179. 54

The candy bar usually sells at a price of two for 90 cents, which is 45 cents per candy bar (because $90 \div 2 = 45$). The sale price is three for $1.05, which is 35 cents per candy bar (because $105 \div 3 = 35$). Thus, Heidi saved 10 cents on every candy bar she bought. She saved $5.40, so she bought 54 candy bars (because $540 \div 10 = 54$).

180. 15 days

Make a list of how much money you will have saved at the end of each day:

1 day: $1

2 days: $1 + $2 = $3

3 days: $3 + $4 = $7

4 days: $7 + $8 = $15

5 days: $15 + $16 = $31

Notice the pattern: Multiply the number 2 by itself as many times as the number of days, and then subtract 1. For example, to find the amount for the 6th day:

$$2 \times 2 \times 2 \times 2 \times 2 \times 2 - 1$$
$$= 2^6 - 1$$
$$= 64 - 1 = 63$$

Thus, the sixth day is $63. These numbers increase very quickly. For example, here is the amount for the 10th day:

$$2^{10} - 1$$
$$= 1,024 - 1 = 1,023$$

And here is the amount for the 14th day:

$$2^{14} - 1$$
$$= 16,384 - 1 = 16,383$$

And here is the amount for the 15th day:

$$2^{15} - 1$$
$$= 32,678 - 1 = 32,677$$

---

181. **i. Yes, ii. Yes, iii. No, iv. No, v. Yes**

A number is divisible by 2 if and only if it's an even number — that is, its last digit is 2, 4, 6, 8, or 0.

i. 32 ends in 2, which is even, so it's divisible by 2.

ii. 70 ends in 0, which is even, so it's divisible by 2.

iii. 109 ends in 9, which is odd, so it's not divisible by 2.

iv. 8,645 ends in 5, which is odd, so it's not divisible by 2.

v. 231,996 ends in 6, which is even, so it's divisible by 2.

---

182. **i. Yes, ii. No, iii. Yes, iv. Yes, v. No**

A number is divisible by 3 if and only if its digits add up to a number that's divisible by 3.

i. The digits in 51 add up to 5 + 1 = 6, which is divisible by 3, so 51 is divisible by 3.

ii. The digits in 77 add up to 7 + 7 = 14, which isn't divisible by 3, so 77 isn't divisible by 3.

iii. The digits in 138 add up to 1 + 3 + 8 = 12, which is divisible by 3, so 138 is divisible by 3.

iv. The digits in 1,998 add up to 1 + 9 + 9 + 8 = 27, which is divisible by 3, so 1,998 is divisible by 3.

v. The digits in 100,111 add up to 1 + 0 + 0 + 1 + 1 + 1 = 4, which isn't divisible by 3, so 100,111 isn't divisible by 3.

183.
i. No, ii. Yes, iii. Yes, iv. Yes, v. No

A number is divisible by 4 if and only if it's an even number whose last two digits form a number that's divisible by 4.

i. 57 is an odd number, so 57 isn't divisible by 4.

ii. The last two digits of 552 are 52, which is divisible by 4 ($52 \div 4 = 13$), so 552 is divisible by 4.

iii. The last two digits of 904 are 04, which is divisible by 4 ($4 \div 4 = 1$), so 904 is divisible by 4.

iv. The last two digits of 12,332 are 32, which is divisible by 4 ($32 \div 4 = 8$), so 12,332 is divisible by 4.

v. The last two digits of 7,435,830 are 30, which isn't divisible by 4 ($30 \div 4 = 7r2$), so 7,435,830 isn't divisible by 4.

184.
i. Yes, ii. No, iii. Yes, iv. Yes, v. No

A number is divisible by 5 if and only if its last digit is 5 or 0.

i. 190 ends in 0, so it's divisible by 5.

ii. 723 ends in 3, so it's not divisible by 5.

iii. 1,005 ends in 5, so it's divisible by 5.

iv. 252,525 ends in 5, so it's divisible by 5.

v. 505,009 ends in 9, so it's not divisible by 5.

185.
i. No, ii. No, iii. Yes, iv. Yes, v. Yes

A number is divisible by 6 if and only if it is divisible by both 2 and 3 — that is, if it's an even number whose digits add up to a number that's divisible by 3.

i. 61 is an odd number, so it isn't divisible by 6.

ii. 88 is an even number whose digits add up to $8 + 8 = 16$, which isn't divisible by 3, so 88 isn't divisible by 6.

iii. 372 is an even number whose digits add up to $3 + 7 + 2 = 12$, which is divisible by 3, so 372 is divisible by 6.

iv. 8,004 is an even number whose digits add up to $8 + 0 + 0 + 4 = 12$, which is divisible by 3, so 8,004 is divisible by 6.

v. 1,001,010 is an even number whose digits add up to $1 + 0 + 0 + 1 + 0 + 1 + 0 = 3$, which is divisible by 3, so 1,001,010 is divisible by 6.

186.
i. No, ii. No, iii. Yes, iv. No, v. Yes

A number is divisible by 8 if and only if it's an even number that's divisible by 4 whose last three digits form a number that's divisible by 8.

i. 881 is an odd number, so it isn't divisible by 8.

ii. The last three digits of 1,914 are 914, which isn't divisible by 8, so 1,914 isn't divisible by 8.

iii. The last three digits of 39,888 are 888, which is divisible by 8 (888 ÷ 8 = 111), so 39,888 is divisible by 8.

iv. The last three digits of 711,124 are 124, which isn't divisible by 8 (124 ÷ 8 = 15r4), so 711,124 isn't divisible by 8.

v. The last three digits of 43,729,408 are 408, which is divisible by 8 (408 ÷ 8 = 51), so 43,729,408 is divisible by 8.

187.

i. No, ii. Yes, iii. No, iv. No, v. Yes

A number is divisible by 9 if and only if its digits add up to a number that is divisible by 9.

i. The digits in 98 add up to 9 + 8 = 17, which isn't divisible by 9, so 98 isn't divisible by 9.

ii. The digits in 324 add up to 3 + 2 + 4 = 9, which is divisible by 9, so 324 is divisible by 9.

iii. The digits in 6,009 add up to 6 + 0 + 0 + 9 = 15, which isn't divisible by 9, so 6,009 isn't divisible by 9.

iv. The digits in 54,321 add up to 5 + 4 + 3 + 2 + 1 = 15, which isn't divisible by 9, so 54,321 isn't divisible by 9.

v. The digits in 993,996 add up to 9 + 9 + 3 + 9 + 9 + 6 = 45, which is divisible by 9, so 993,996 is divisible by 9.

188.

i. Yes, ii. No, iii. No, iv. Yes, v. Yes

A number is divisible by 10 if and only if it ends with a 0.

i. 340 ends with a 0, so 340 is divisible by 10.

ii. 8,245 ends with a 5, so 8,245 isn't divisible by 10.

iii. 54,002 ends with a 2, so 54,002 isn't divisible by 10.

iv. 600,010 ends with a 0, so 600,010 is divisible by 10.

v. 1,010,100 ends with a 0, so 1,010,100 is divisible by 10.

189.

i. No, ii. Yes, iii. No, iv. Yes, v. Yes

A number is divisible by 11 if and only if the alternating difference and sum of its digits results in a number that is divisible by 11 (including 0 and negative numbers).

i. 1 − 3 + 4 = 2, which isn't divisible by 11, so 134 isn't divisible by 11.

ii. 2 − 0 + 9 = 11, which is divisible by 11, so 209 is divisible by 11.

iii. 6 − 8 + 1 = −1, which isn't divisible by 11, so 681 isn't divisible by 11.

iv. 1 − 9 + 2 − 5 = −11, which is divisible by 11, so 1,925 is divisible by 11.

v. 8 − 1 + 9 − 2 + 8 = 22, which is divisible by 11, so 81,928 is divisible by 11.

190.

i. No, ii. Yes, iii. No, iv. Yes, v. No

A number is divisible by 12 if and only if it's divisible by both 3 and 4 — that is, if its digits add up to a number that's divisible by 3, and if it's also an even number whose last two digits form a number that's divisible by 4.

i. 81 is an odd number, so it isn't divisible by 4; therefore it isn't divisible by 12.

ii. The sum of the digits in 132 is 1 + 3 + 2 = 6, which is divisible by 3, so 132 is divisible by 3. And the last two digits of 132 are 32, which is divisible by 4 (32 ÷ 4 = 8), so 132 is also divisible by 4. Therefore, 132 is divisible by 12.

iii. The sum of the digits in 616 is 6 + 1 + 6 = 13, which isn't divisible by 3, so 616 isn't divisible by 12. Therefore, it isn't divisible by 12.

iv. The last two digits of 123,456 are 56, which is divisible by 4 (56 ÷ 4 = 14), so 123,456 is divisible by 4. The sum of the digits in 123,456 is 1 + 2 + 3 + 4 + 5 + 6 = 21, which is divisible by 3, so 123,456 is also divisible by 3. Therefore, 123,456 is divisible by 12.

v. The last two digits of 12,345,678 are 78, which isn't divisible by 4 (78 ÷ 4 = 19r2), so 12,345,678 isn't divisible by 4. Therefore, it isn't divisible by 12.

191. 1,000

The number 87,000 ends with 3 zeros, so its greatest factor that is a power of 10 also has 3 zeros. Therefore, this number is 1,000.

192. 100,000

The number 9,200,000 ends with 5 zeros, so its greatest factor that is a power of 10 also has 5 zeros. Therefore, this number is 100,000.

193. 10

The number 30,940,050 ends with 1 zero, so its greatest factor that is a power of 10 also has 1 zero. Therefore, this number is 10.

194. 2, 3, and 6

The number 78 is even, so it's divisible by 2.

Its digits add up to 7 + 8 = 15, which is divisible by 3, so it's divisible by 3.

Its last two digits form the number 78, which isn't divisible by 4, so it isn't divisible by 4.

It ends in 8, so it isn't divisible by 5.

And it's divisible by both 2 and 3, so it's also divisible by 6.

195. 2 and 4

The number 128 is even, so it's divisible by 2.

Its digits add up to 1 + 2 + 8 = 11, which isn't divisible by 3, so it isn't divisible by 3 or 6.

Its last two digits form the number 28, which is divisible by 4, so it's divisible by 4.

And it ends in 8, so it isn't divisible by 5.

196. **2, 4, and 5**

The number 380 is even, so it's divisible by 2.

Its digits add up to $3 + 8 + 0 = 11$, which isn't divisible by 3, so it isn't divisible by 3 or 6.

Its last two digits form the number 80, which is divisible by 4, so it's divisible by 4.

And it ends in 0, so it's divisible by 5.

197. **3 and 5**

The number 6,915 is odd, so it isn't divisible by 2, 4, or 6.

Its digits add up to $6 + 9 + 1 + 5 = 21$, which is divisible by 3, so it's divisible by 3.

And it ends in 5, so it's divisible by 5.

198. **3**

The number 59 is 3 greater than 56. So when you divide $59 \div 7$, the remainder is 3.

199. **8**

The number 611 is 1 less than 612. So when you divide $611 \div 9$, the remainder is just 1 less than 9, so the remainder is 8.

200. **1**

The number 8,995 is 5 less than 9,000. So when you divide $8,995 \div 6$, the remainder is just 5 less than 6, so the remainder is 1.

201. **i. composite, ii. prime, iii. composite, iv. prime, v. composite**

Any number less than 121 (because $11 \times 11 = 121$) that isn't divisible by 2, 3, 5, or 7 is a prime number.

i. $3 + 9 = 12$, which is divisible by 3, so 39 is a composite number.

ii. 41 ends in 1, so it isn't divisible by 2 or 5. And $4 + 1 = 5$, which isn't divisible by 3, so 41 isn't divisible by 3. Finally, $41 \div 7 = 5r6$, so 41 isn't divisible by 7. Therefore, 41 is a prime number.

iii. $5 + 7 = 12$, which is divisible by 3, so 57 is a composite number.

iv. 73 ends in 3, so it isn't divisible by 2 or 5. And $7 + 3 = 10$, which isn't divisible by 3, so 73 isn't divisible by 3. Finally, $73 \div 7 = 10r3$, so 73 isn't divisible by 7. Therefore, 73 is a prime number.

v. $91 \div 7 = 13$, so 91 is divisible by 7. Therefore, 91 is a composite number.

202.

No

Any number less than 169 (because $13 \times 13 = 169$) that isn't divisible by 2, 3, 5, 7, or 11 is a prime number. But $1 - 4 + 3 = 0$, which is divisible by 11, so 143 is divisible by 11. Therefore, 143 isn't a prime number, so the answer is No.

203.

Yes

Any number less than 169 (because $13 \times 13 = 169$) that isn't divisible by 2, 3, 5, 7, or 11 is a prime number. The number 151 ends in 1, so it isn't divisible by 2 or 5. Its digits add to $1 + 5 + 1 = 7$, which isn't divisible by 3, so it isn't divisible by 3. It isn't divisible by 7 ($151 \div 7 = 21r4$). And $1 - 5 + 1 = 3$, which isn't divisible by 11, so 151 isn't divisible by 11. Therefore, 151 is a prime number, so the answer is Yes.

204.

No

Any number less than 169 (because $13 \times 13 = 169$) that isn't divisible by 2, 3, 5, 7, or 11 is a prime number. However, 161 is divisible by 7 ($161 \div 7 = 23$), so it isn't a prime number. Therefore, the answer is No.

205.

Yes

Any number less than 289 (because $17 \times 17 = 289$) that isn't divisible by 2, 3, 5, 7, 11, or 13 is a prime number. The number 223 ends in 3, so it isn't divisible by 2 or 5. It's digits add up to $2 + 2 + 3 = 8$, which isn't divisible by 3, so it isn't divisible by 3. $2 - 2 + 3 = 3$, which isn't divisible by 11, so 223 isn't divisible by 11. Finally, test 223 for divisibility by 7 and 13:

$$223 \div 7 = 31r6$$

$$223 \div 13 = 17r2$$

Therefore, because 267 isn't divisible by 7 or 13, it's a prime number, so the answer is Yes.

206.

No

Any number less than 289 (because $17 \times 17 = 289$) that isn't divisible by 2, 3, 5, 7, 11, or 13 is a prime number. However, $2 + 6 + 7 = 15$, which is divisible by 3, so 267 is divisible by 3. Therefore, 267 isn't a prime number, so the answer is No.

207.

3 and 31

The sum of the digits in 93 is $9 + 3 = 12$, which is divisible by 3, so 93 is divisible by 3. And $93 \div 3 = 31$, which is also a prime number. So 93 is divisible by both 3 and 31.

208.

3 and 11

The sum of the digits in 297 is $2 + 9 + 7 = 18$, which is divisible by 3, so 297 is divisible by 3. And $2 - 9 + 7 = 0$, so 297 is divisible by 11.

209.

2 and 7

The number 448 is even, so it's divisible by 2. And $448 \div 2 = 224$, which is also even, so divide by 2 again: $224 \div 2 = 112$. This, too, is even, so divide by 2 again: $112 \div 2 = 56$. At this point, you may notice that 56 is divisible by 7 ($56 \div 7 = 8$), so 448 is divisible by both 2 and 7.

210.

5, 11, and 97

The number 293,425 ends in 5, so it's divisible by 5, which is a prime number. And $293,425 \div 5 = 56,685$, which is also divisible by 5, so you can divide by 5 again: $56,685 \div 5 = 11,737$. Now, notice that $1 - 1 + 7 - 3 + 7 = 11$, so 11,737 is divisible by 11. Thus, 293,425 is also divisible by 11, which is also a prime number. And $11,737 \div 11 = 1,067$. Now, notice that $1 - 0 + 6 - 7 = 0$, so 1,067 is also divisible by 11. And $1,067 \div 11 = 97$. Finally, 97 is a prime number (because it's less than 121 and not divisible by 2, 3, 5, or 7). Therefore, 293,425 is divisible by 5, 11, and 97.

211.

78 and 3,000

Only even numbers (numbers whose last digit is 2, 4, 6, 8, or 0) have a factor of 2. Therefore, 78 and 3,000 both have a factor of 2, but 181, 222,225, and 1,234,569 don't have a factor of 2.

212.

3,000 and 222,225

Only numbers whose last digit is 5 or 0 have a factor of 5. Therefore, 3,000 and 222,225 both have a factor of 5, but 78, 181, and 1,234,569 don't have a factor of 5.

213.

78; 3,000; 222,225; and 1,234,569

The only numbers that are divisible by 3 (and therefore have a factor of 3) are those whose digits add up to a number that is also divisible by 3. Test all five numbers as follows:

$$7 + 8 = 15$$

$$1 + 8 + 1 = 10$$

$$3 + 0 + 0 + 0 = 3$$

$$2 + 2 + 2 + 2 + 5 = 15$$

$$1 + 2 + 3 + 4 + 5 + 6 + 9 = 30$$

Therefore, 78, 3,000, 222,225, and 1,234,569 all have a factor of 3, but 181 doesn't have a factor of 3.

214.

3,000

Only numbers whose last digit is 0 have a factor of 10. Therefore, 3,000 has a factor of 10, but 78, 181, 222,225, and 1,234,569 don't have a factor of 10.

215. 1,234,569

$$78 \div 7 = 11r1$$

$$181 \div 7 = 25r6$$

$$3,000 \div 7 = 428r4$$

$$222,225 \div 7 = 31,746r3$$

$$1,234,569 \div 7 = 176,367$$

Therefore, 1,234,569 has a factor of 7, but 78, 181, 3,000, and 222,225 don't have a factor of 7.

216. 6

Begin by writing the numbers 1 and 12, with plenty of space between them:

Factors of 12: 1,   12

The number 12 is even, so 2 is a factor, and so is 6 (because $12 \div 2 = 6$):

Factors of 12: 1, 2,   6, 12

The number 12 is divisible by 3, and also 4 (because $12 \div 3 = 4$):

Factors of 12: 1, 2, 3, 4, 6, 12

Counting these factors, you find there are six different factors of 12.

217. 3

Begin by writing the numbers 1 and 25, with plenty of space between them:

Factors of 25: 1,   25

The number 25 isn't even, so 2 isn't a factor. It also isn't divisible by 3 (because 2 + 5 = 7, which isn't divisible by 3). It is divisible by 5 (because $25 \div 5 = 5$), so:

Factors of 25: 1, 5, 25

Counting these factors, you find there are three different factors of 25.

218. 6

Begin by writing the numbers 1 and 32, with plenty of space between them:

Factors of 32: 1   32

The number 32 is even, so 2 is a factor, and so is 16 (because $32 \div 2 = 16$):

Factors of 32: 1, 2,   16, 32

The number 32 isn't divisible by 3 (because 3 + 2 = 5, which isn't divisible by 3). It is divisible by 4, and also 8 (because $32 \div 4 = 8$), so:

Factors of 32: 1, 2, 4, 8, 16, 32

The number 32 isn't divisible by 6 or 7, so the list above is complete. Counting these factors, you find there are six different factors of 32.

219. 4

Begin by writing the numbers 1 and 39, with plenty of space between them:

Factors of 39: 1,   39

The number 39 isn't even, so 2 isn't a factor. It is divisible by 3 (because 3 + 9 = 12, which is divisible by 3), and also 13 (because 39 ÷ 3 = 13), so:

Factors of 39: 1, 3,   13, 39

The number 39 isn't divisible by 5, because its last digit isn't 5 or 0. It also isn't divisible by 7, because 39 ÷ 7 = 5r4. Therefore, these are the only factors of 39:

Factors of 39: 1, 3, 13, 39

Counting these factors, you find there are four different factors of 39.

220. 2

Begin by writing the numbers 1 and 41, with plenty of space between them:

Factors of 41: 1,   41

The number 41 isn't even, so 2 isn't a factor. It isn't divisible by 3, because 4 + 1 = 5 (its digits don't add up to a number divisible by 3). It isn't divisible by 5, because its last digit isn't 5 or 0. And it isn't divisible by 7, because 41 ÷ 7 = 5r6. Therefore, 41 is prime, so its only factors are 1 and 41:

Factors of 41: 1, 41

221. 6

Begin by writing the numbers 1 and 63, with plenty of space between them:

Factors of 63: 1,   63

The number 63 isn't even, so 2 isn't a factor. It is divisible by 3, because 63 ÷ 3 = 21, so:

Factors of 63: 1, 3,   21, 63

It isn't divisible by 5, because its last digit isn't 5 or 0. It is divisible by 7, because 63 ÷ 7 = 9, so:

Factors of 63: 1, 3, 7, 9, 21, 63

It isn't divisible by 8 because it isn't an even number, so the preceding list is complete. Counting these factors, you find there are six different factors of 63.

222. 12

Begin by writing the numbers 1 and 90, with plenty of space between them:

Factors of 90: 1,   90

The number 90 is even, so 2 is a factor, and so is 45 (because 90 ÷ 2 = 45), so:

Factors of 90: 1, 2,   45, 90

It is divisible by 3, because $90 \div 3 = 30$, so:

Factors of 90: 1, 2, 3,   30, 45, 90

It isn't divisible by 4, because $90 \div 4 = 22r2$. It is divisible by 5, $90 \div 5 = 18$, so:

Factors of 90: 1, 2, 3, 5,   18, 30, 45, 90

It is divisible by 6, because $90 \div 6 = 15$, so:

Factors of 90: 1, 2, 3, 5, 6,   15, 18, 30, 45, 90

It isn't divisible by 7 or 8, but it is divisible by 9, because $90 \div 9 = 10$, so:

Factors of 90: 1, 2, 3, 5, 6, 9, 10, 15, 18, 30, 45, 90

Counting these factors, you find there are twelve different factors of 90.

223.   16

Begin by writing the numbers 1 and 120, with plenty of space between them:

Factors of 120: 1,   120

The number 120 is divisible by 2, 3, 4, 5, and 6, as follows:

$$120 \div 2 = 60$$
$$120 \div 3 = 40$$
$$120 \div 4 = 30$$
$$120 \div 5 = 24$$
$$120 \div 6 = 20$$

Therefore:

Factors of 120: 1, 2, 3, 4, 5, 6,   20, 24, 30, 40, 60, 120

It isn't divisible by 7, because $120 \div 7 = 17r1$. It is divisible by 8, because $120 \div 8 = 15$. Therefore:

Factors of 120: 1, 2, 3, 4, 5, 6, 8, 10, 12, 15, 20, 24, 30, 40, 60, 120

It isn't divisible by 9, because $120 \div 9 = 13r3$.

Counting these factors, you find there are 16 different factors of 120.

224.   6

Before you begin, notice that 171 is less than 196, and that $14 \times 14 = 196$, so you only need to check the numbers below 13 and find the numbers that the larger factors are paired with.

Begin by writing the numbers 1 and 171, with plenty of space between them:

Factors of 171: 1,   171

The number 171 isn't even, so 2 isn't a factor. It is divisible by 3 (because $1 + 7 + 1 = 9$, which is divisible by 3), so divide: $171 \div 3 = 57$. Therefore:

Factors of 171: 1, 3,   57, 171

It isn't divisible by 5, because its last digit isn't 5 or 0. And it isn't divisible by 7, because $171 \div 7 = 24r3$. It is divisible by 9 (because $1 + 7 + 1 = 9$, which is divisible by 9), so divide $171 \div 9 = 19$. Therefore:

Factors of 171: 1, 3, 9, 19, 57, 171

It isn't divisible by 10, 11, 12, or 13, so the preceding list is complete. Counting these factors, you find there are six different factors of 171.

225. 16

Before you begin, notice the following:
$$1,000 = 10 \times 10 \times 10 = 2 \times 5 \times 2 \times 5 \times 2 \times 5$$

Thus, the number 1,000 has no factors that are multiples of any prime numbers other than 2 or 5.

Begin by writing the numbers 1 and 1,000, with plenty of space between them:

Factors of 1,000: 1,   1,000

Now, test all possible numbers from 2 to 10:

$$1,000 \div 2 = 500$$
$$1,000 \div 4 = 250$$
$$1,000 \div 5 = 200$$
$$1,000 \div 8 = 125$$
$$1,000 \div 10 = 100$$

Therefore:

Factors of 1,000: 1, 2, 4, 5, 8, 10,   100, 125, 200, 250, 500, 1,000

It isn't divisible by 16, because $1,000 \div 16 = 62r8$.

It is divisible by 20, because $1,000 \div 20 = 50$, so:

Factors of 1,000: 1, 2, 4, 5, 8, 10, 20, 50, 100, 125, 200, 250, 500, 1,000

It is divisible by 25, because $1,000 \div 25 = 40$, so:

Factors of 1,000: 1, 2, 4, 5, 8, 10, 20, 25, 40, 50, 100, 125, 200, 250, 500, 1,000

The preceding list is complete. Counting these factors, you find there are 16 different factors of 1,000.

226. 3

Decompose 30 into its prime factors using a factor tree. Here's one possible factoring tree:

Thus, $30 = 2 \times 3 \times 5$, so 30 has 3 nondistinct prime factors.

227. 3

Decompose 66 into its prime factors using a factor tree. Here's one possible factoring tree:

Thus, $66 = 2 \times 3 \times 11$, so 66 has 3 nondistinct prime factors.

228. 4

Decompose 81 into its prime factors using a factor tree. Here's one possible factoring tree:

Thus, $81 = 3 \times 3 \times 3 \times 3$, so 81 has 4 nondistinct prime factors.

229. 1

The number 97 isn't even, so it isn't divisible by 2. It doesn't end with 5 or 0, so it isn't divisible by 5. Its digits don't add up to a multiple of 3 (9 + 7 = 16), so it isn't divisible by 3. Finally, it isn't divisible by 7 (97 ÷ 7 = 13*r*6).

Thus, 97 is a 2-digit number that isn't divisible by 2, 3, 5, or 7, so it's a prime number. Therefore, it cannot be decomposed into smaller prime numbers.

230. 3

Decompose 98 into its prime factors using a factor tree. Here's one possible factoring tree:

Thus, $98 = 2 \times 7 \times 7$, so 98 has 3 nondistinct prime factors.

231. 6

Decompose 216 into its prime factors using a factor tree. Here's one possible factoring tree:

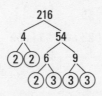

Thus, $216 = 2 \times 2 \times 2 \times 3 \times 3 \times 3$, so 216 has 6 nondistinct prime factors.

232. 7

Decompose 800 into its prime factors using a factor tree. Here's one possible factoring tree:

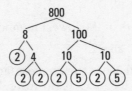

Thus, $800 = 2 \times 2 \times 2 \times 2 \times 2 \times 5 \times 5$, so 800 has 7 nondistinct prime factors.

233. 4

Generate the factors of both 16 and 20:

Factors of 16: 1, 2, 4, 8, 16

Factors of 20: 1, 2, 4, 5, 10, 20

Therefore, the GCF of 16 and 20 is 4.

234. 6

Generate the factors of both 12 and 30:

Factors of 12: 1, 2, 4, 6, 12

Factors of 30: 1, 2, 3, 5, 6, 10, 15, 30

Therefore, the GCF of 12 and 30 is 6.

235. 5

Generate the factors of both 25 and 55:

Factors of 25: 1, 5, 25

Factors of 55: 1, 5, 11, 55

Therefore, the GCF of 25 and 55 is 5.

236. 26

Generate the factors of both 26 and 78:

Factors of 26: 1, 2, 13, 26

Factors of 78: 1, 2, 3, 6, 13, 26, 39, 78

Therefore, the GCF of 26 and 78 is 26.

237. 25

Generate the factors of the lowest number, which is 125:

Factors of 125: 1, 5, 25, 125

The GCF of 125 and 350 is the greatest factor of 125 that is also a factor of 350. It isn't 125, because $350 \div 125 = 2r100$. However, $350 \div 25 = 14$, so 25 is the GCF of 125 and 350.

238. 1

Generate the factors of the lowest number, which is 28:

Factors of 28: 1, 2, 4, 7, 14, 28

The GCF of 28, 35, and 48 is the greatest factor of 28 that is also a factor of 35 and 48 . It isn't an even number, because 35 isn't even. It isn't 7, because $48 \div 7 = 6r6$. Therefore, the GCF of 28, 35, and 48 is 1.

239. 3

Generate the factors of the lowest number, which is 18:

Factors of 18: 1, 2, 3, 6, 9, 18

The GCF of 18, 30, and 99 is the greatest factor of 18 that is also a factor of 30 and 99. It isn't an even number, because 99 isn't even. It isn't 9, because $30 \div 9 = 3r3$. However, $30 \div 3 = 10$ and $99 \div 3 = 33$, so 3 is the GCF of 18, 30, and 99.

240. 11

Generate the factors of the lowest number, which is 33:

Factors of 33: 1, 3, 11, 33

The GCF of 33, 77, and 121 is the greatest factor of 33 that is also a factor of 77 and 121. It isn't 33, because $77 \div 33 = 2r11$. However, $33 \div 11 = 3$ and $121 \div 11 = 11$, so 11 is the GCF of 33, 77, and 121.

241. **20**

Generate the factors of the lowest number, which is 40:

Factors of 40: 1, 2, 4, 5, 8, 10, 20, 40

The GCF of 40, 60, and 220 is the greatest factor of 40 that is also a factor of 60 and 220. It isn't 40, because $60 \div 40 = 1r20$. However, $60 \div 20 = 3$ and $220 \div 20 = 11$, so 20 is the GCF of 40, 60, and 220.

242. **18**

Generate the factors of the lowest number, which is 90:

Factors of 90: 1, 2, 3, 5, 6, 9, 10, 15, 18, 30, 45, 90

The GCF of 90, 126, 180, and 990 is the greatest factor of 90 that is also a factor of 126, 180, and 990. It isn't a multiple of 5, because 126 isn't divisible by 5. So that rules out 90, 45, and 30 from the top of the list. However, $126 \div 18 = 7$, $180 \div 18 = 10$, and $990 \div 18 = 55$, so 18 is the GCF of 90, 126, 180, and 990.

243. **7**

Generate the multiples of 4 as follows:

Multiples of 4: 4, 8, 12, 16, 20, 24, 28

Therefore, 4 has 7 multiples between 1 and 30.

244. **11**

Generate the multiples of 6 as follows:

Multiples of 6: 6, 12, 18, 24, 30, 36, 42, 48, 54, 60, 66

Therefore, 6 has 11 multiples between 1 and 70.

245. **14**

Generate the multiples of 7 as follows:

Multiples of 7: 7, 14, 21, 28, 35, 42, 49, 56, 63, 70, 77, 84, 91, 98

Therefore, 7 has 14 multiples between 1 and 100.

246. **12**

Generate the multiples of 12 as follows:

Multiples of 12: 12, 24, 36, 48, 60, 72, 84, 96, 108, 120, 132, 144

Therefore, 12 has 12 multiples between 1 and 150.

247. 11

Generate the multiples of 15 as follows:

Multiples of 15: 15, 30, 45, 60, 75, 90, 105, 120, 135, 150, 165

Therefore, 15 has 11 multiples between 1 and 175.

248. 12

Generate the multiples of 16 as follows:

Multiples of 16: 16, 32, 48, 64, 80, 96, 112, 128, 144, 160, 176, 192

Therefore, 16 has 12 multiples between 1 and 200.

249. 13

Generate the multiples of 75 as follows:

Multiples of 75: 75, 150, 225, 300, 375, 450, 525, 600, 675, 750, 825, 900, 975

Therefore, 75 has 13 multiples between 1 and 1,000.

250. 24

Generate the multiples of both 6 and 8 until you find the lowest number that appears in both lists:

Multiples of 6: 6, 12, 18, 24

Multiples of 8: 8, 16, 24

Therefore, the LCM of 6 and 8 is 24.

251. 77

The LCM of a set of prime numbers is always the product of those numbers:

$$7 \times 11 = 77$$

252. 28

Generate the multiples of both 4 and 14 until you find the lowest number that appears in both lists:

Multiples of 4: 4, 8, 12, 16, 20, 24, 28

Multiples of 14: 14, 28

Therefore, the LCM of 4 and 14 is 28.

253. 60

Generate the multiples of both 12 and 15 until you find the lowest number that appears in both lists:

Multiples of 12: 12, 24, 36, 48, 60

Multiples of 15: 15, 30, 45, 60

Therefore, the LCM of 12 and 15 is 60.

254. 72

Generate the multiples of both 8 and 18 until you find the lowest number that appears in both lists:

Multiples of 8: 8, 16, 24, 32, 40, 48, 56, 64, 72

Multiples of 18: 18, 36, 54, 72

Therefore, the LCM of 8 and 18 is 72.

255. 180

Generate the multiples of both 20 and 45 until you find the lowest number that appears in both lists:

Multiples of 20: 20, 40, 60, 80, 100, 120, 140, 160, 180

Multiples of 45: 45, 90, 135, 180

Therefore, the LCM of 20 and 45 is 180.

256. 80

To begin, notice that 8 is a factor of 16, so every multiple of 16 is also a multiple of 8. Additionally, all multiples of 10 end in 0, so you don't have to generate this list. Thus, generate the multiples of 16 until you find a number that ends in 0:

Multiples of 16: 16, 32, 48, 64, 80

Therefore, the LCM of 8, 10, and 16 is 80.

257. 36

To begin, notice that 4 is a factor of 12, so every multiple of 12 is also a multiple of 4. So you only need to generate the multiples of both 12 and 18 until you find the lowest number that appears in both lists:

Multiples of 12: 12, 24, 36

Multiples of 18: 18, 36

Therefore, the LCM of 4, 12, and 18 is 36.

258. 357

The LCM of a set of prime numbers is always the product of those numbers:

$$3 \times 7 \times 17 = 357$$

259. 840

For difficult LCM problems, generate the prime factors of each number.

$$10 = 2 \times 5$$
$$14 = 2 \times 7$$
$$24 = 2 \times 2 \times 2 \times 3$$

Now, find the greatest number of occurrences for each prime in any number in the list: There are three factors of 2 (in 24), one factor of 3 (in 24), one factor of 5 (in 10), and one factor of 7 (in 14). Multiply these numbers:

$$2 \times 2 \times 2 \times 3 \times 5 \times 7 = 840$$

Therefore, the LCM of 10, 14, and 24 is 840.

260. 13,200

For difficult LCM problems, generate the prime factors of each number.

$$11 = 11$$
$$15 = 3 \times 5$$
$$16 = 2 \times 2 \times 2 \times 2$$
$$25 = 5 \times 5$$

Now, find the greatest number of occurrences for each prime in any number in the list: There are four factors of 2 (in 16), one factor of 3 (in 15), two factors of 5 (in 25), and one factor of 11 (in 11). Multiply these numbers:

$$2 \times 2 \times 2 \times 2 \times 3 \times 5 \times 5 \times 11 = 13,200$$

Therefore, the LCM of 10, 14, and 24 is 13,200.

261. 21

The class broke into groups of 3 and 7, so the number of children in this class is divisible by both 3 and 7. This number is also less than 40. The only such number is 21 ($3 \times 7 = 21$).

262. 3

The only factors of 57, besides 1 and 57, are 3 and 19. Thus, they transported 3 cats in each of 19 cages.

263. 7

The only factors of 91, besides 1 and 91, are 7 and 13. There were more than 8 tables, so 13 tables accommodated 7 guests each.

264.   5

To begin, find all of the factors of 105:

Factors of 105: 1, 3, 5, 7, 15, 21, 35, 105

Thus, the only factor of 105 between 20 and 30 is 21, so each child received exactly 21 pieces. Therefore, because $105 \div 5 = 21$, Mary Ann has 5 children.

265.   6

The prime factorization of $132 = 2 \times 2 \times 3 \times 11$. Thus, 132 isn't divisible by 5, 7, 8, or 9. However, $132 \div 6 = 22$.

266.   14 and 15

Begin by finding the prime factorization of $210 = 2 \times 3 \times 5 \times 7$. The only two numbers between 10 and 20 that are factors of these numbers are $2 \times 7 = 14$ and $3 \times 5 = 15$.

267.   56 days

Maxine must complete her inspection again in 8 days, 16 days, and so forth. Norma must complete hers in 14 days, 28 days, and so forth. To find out the next day when they will both complete their inspections, generate the multiples of 8 and 14:

Multiples of 8: 8, 16, 24, 32, 40, 48, 56

Multiples of 14: 14, 28, 42, 56

Therefore, they will both complete their inspection again in 56 days.

268.   11 feet

The dimensions of the room are all whole numbers of feet, so the height of the room must be a factor of 2,816. To begin, find the prime factorization of 2,816:

$$2,816 = 2 \times 2 \times 2 \times 2 \times 2 \times 2 \times 2 \times 2 \times 11$$

The only odd factor of 2,816 is 11, so the height of the room is 11 feet.

269.   60

Begin by writing down the prime factorizations of the numbers 3, 4, 5, and 6:

$$3 = 3$$
$$4 = 2 \times 2$$
$$5 = 5$$
$$6 = 2 \times 3$$

Among these four prime factorizations, 2 shows up no more than two times in any case, 3 shows up no more than once in any case, and 5 shows up no more than once in any case. Thus, multiply two 2s, one 3, and one 5:

$$2 \times 2 \times 3 \times 5 = 60$$

Therefore, a group of 60 people is the smallest group of people that can be divided into subgroups of three, four, five, or six people.

270. 7

Marion was able to divide all but 2 of the 100 apples evenly, so she was able to divide 98 apples evenly. The factors of 98 are as follows:

Factors of 98: 1, 2, 7, 14, 49, 98.

The group included fewer than 12 people, so the only remaining options are 2 or 7. If the group had included only 2 people, Marion could have given 50 apples to each person in the group with no remaining apples. Therefore, the group included 7 people.

271. 36

Any number divisible by 4 must be an even number. So, begin by writing down the first few even square numbers: 4, 16, 36, 64, 100, 144…

All of these are divisible by 4. But 4 isn't divisible by 3, and neither is 16. However, 36 is divisible by 3.

272. 7 meters

The two sides of the 168-square-meter ballroom are whole numbers of meters, so begin by finding the factors of 168:

Factors of 168: 1, 2, 3, 4, 6, 7, 8, 12, 14, 21, 24, 28, 42, 56, 84, 168

Thus, the longer side of the room is 24 meters (because that's the only factor between 21 and 28), so the shorter side is 7 meters (because $168 \div 24 = 7$).

273. 63

The original group contained from 50 to 100 people, and is evenly divisible by 21. The only two possible numbers are 63 and 84. But, the group could not be paired up evenly, so it contains an odd number of people. Therefore, it contains 63 people.

274. 77

To begin, generate the multiples of 7, starting with the numbers greater than 50: 56, 63, 70, 77, …

The numbers 56 and 70 are both divisible by 2. The number 63 is divisible by 3 (because 6 + 3 = 9, which is divisible by 3). However, the number 77 is an odd number, so it isn't divisible by 2, 4, or 6. It ends in 7, so it isn't divisible by 5. And the sum of its digits is 7 + 7 = 14, which isn't divisible by 3, so it isn't divisible by 3.

275. 31

To solve this problem, first find the lowest number that is divisible by 2, 3, and 5. Because these three numbers are prime, simply multiply $2 \times 3 \times 5 = 30$. Now, add 1 for the one person who was always left out of the groups:

$$30 + 1 = 31$$

Check this result by dividing:

$$31 \div 2 = 15r1$$

$$31 \div 3 = 10r1$$

$$31 \div 5 = 6r1$$

Therefore, the group contained 31 people.

276. 8

To begin, find the prime factorization of 1,260:

$$1,260 = 2 \times 2 \times 3 \times 3 \times 5 \times 7$$

Thus, it isn't divisible by 8 (because $8 = 2 \times 2 \times 2$).

277. 3

The package contained between 70 and 80 stickers, and this number was divisible by 9. Thus, the package contained exactly 72 stickers, so find the factors of 72:

Factors of 72: 1, 2, 3, 4, 6, 8, 9, 12, 18, 24, 36, 72

When the additional children arrived, Maxwell was able to divide these 72 stickers equally. Thus, there were 12 children at the party at this point, so 3 additional children had arrived.

278. 32

The number of students was a square number from 200 to 300, so it was 225, 256, or 289. Of these, 256 is the only number that is divisible by 8. Therefore, calculate the number of groups of eight students as follows:

$$256 \div 8 = 32$$

279. 21

You can solve this problem by writing down all the 2-digit numbers and then crossing out those that are divisible by 2, 3, 5, and 7. To make this quicker, skip the even numbers as well as those numbers ending in 5:

11, 13, 17, 19, 21, 23, 27, 29, 31, 33, 37, 39, 41, 43, 47, 49, 51, 53, 57, 59, 61, 63, 67, 69, 71, 73, 77, 79, 81, 83, 87, 89, 91, 93, 97, 99

Now, cross off the numbers that are divisible by 3 (that is, whose digits add up to a number that is divisible by 3):

11, 13, 17, 19, 23, 29, 31, 37, 41, 43, 47, 49, 53, 59, 61, 67, 71, 73, 77, 79, 83, 89, 91, 97

Finally, cross off the numbers that are divisible by 7:

11, 13, 17, 19, 23, 29, 31, 37, 41, 43, 47, 53, 59, 61, 67, 71, 73, 79, 83, 89, 97

The remaining 21 numbers are all prime.

---

280. **39**

Solve this problem by generating the numbers that result in a remainder of 3 when divided by 4, 5, and 6:

Div by 4, r is 3: 3, 7, 11, 15, 19, 23, 27, 31, 35, 39

Div by 5, r is 4: 4, 9, 14, 19, 24, 29, 34, 39

Div by 6, r is 3: 3, 9, 15, 21, 27, 33, 39

The first number to appear in all three lists is 39.

---

281. **See below.**

i. $\frac{1}{3}$

ii. $\frac{3}{4}$

iii. $\frac{2}{5}$

iv. $\frac{5}{6}$

v. $\frac{7}{12}$

In each case, the number of shaded pieces is the numerator (top number) of the fraction, and the total number of pieces is the denominator (bottom number).

---

282. **See below.**

i. $\frac{1}{4}$: numerator is 1; denominator is 4

ii. $\frac{2}{9}$: numerator is 2; denominator is 9

iii. $\frac{9}{2}$: numerator is 9; denominator is 2

iv. 4: numerator is 4; denominator is 1

v. 0: numerator is 0; denominator is 1

For i-iii, the numerator is the top number of the fraction and the denominator is the bottom number. For iv and v, the numerator is the value of the whole number and the denominator is 1.

283. **See below.**

i. proper

ii. improper

iii. proper

iv. proper

v. improper

An improper fraction is a fraction whose numerator is *more than* its denominator.

284. **See below.**

i. $\frac{3}{1}$

ii. $\frac{10}{1}$

iii. $\frac{250}{1}$

iv. $\frac{2,000}{1}$

v. $\frac{0}{1}$

To change any whole number to a fraction, draw a line under it and place 1 in the denominator.

285. **i. 3, ii. 4, iii. 9, iv. 2, v. 6**

When the numerator (top number) of a fraction is divisible by the denominator (bottom number), you can change the fraction to a whole number by dividing:

i. $6 \div 2 = 3$

ii. $20 \div 5 = 4$

iii. $54 \div 6 = 9$

iv. $100 \div 50 = 2$

v. $150 \div 25 = 6$

286. **See below.**

i. $\frac{7}{2}$

ii. $\frac{3}{5}$

iii. 10

iv. $\frac{1}{6}$

v. $\frac{100}{99}$

To find the reciprocal of a fraction, flip it over, changing the positions of the numerator (top number) and denominator (bottom number).

287. $\dfrac{11}{5}$

Multiply the whole number by the denominator; then add the numerator:

$2 \times 5 + 1 = 11$

Use this result as the numerator and place it over the original denominator:

$\dfrac{11}{5}$

288. $\dfrac{31}{7}$

Multiply the whole number by the denominator; then add the numerator:

$4 \times 7 + 3 = 31$

Use this result as the numerator and place it over the original denominator:

$\dfrac{31}{7}$

289. $\dfrac{73}{12}$

Multiply the whole number by the denominator; then add the numerator:

$6 \times 12 + 1 = 73$

Use this result as the numerator and place it over the original denominator:

$\dfrac{73}{12}$

290. $\dfrac{97}{10}$

Multiply the whole number by the denominator; then add the numerator:

$9 \times 10 + 7 = 97$

Use this result as the numerator and place it over the original denominator:

$\dfrac{97}{10}$

291. $4\dfrac{1}{3}$

Using integer division, divide the numerator by the denominator:

$13 \div 3 = 4r1$

Use this result to build the equivalent mixed number, using the quotient (4) as the whole number and placing the remainder (1) over the original denominator:

$4\dfrac{1}{3}$

292. $15\frac{1}{2}$

Using integer division, divide the numerator by the denominator:

$$31 \div 2 = 15r1$$

Use this result to build the equivalent mixed number, using the quotient (15) as the whole number and placing the remainder (1) over the original denominator (2):

$$15\frac{1}{2}$$

293. $16\frac{3}{5}$

Using integer division, divide the numerator by the denominator:

$$83 \div 5 = 16r3$$

Use this result to build the equivalent mixed number, using the quotient (16) as the whole number and placing the remainder (3) over the original denominator (5):

$$16\frac{3}{5}$$

294. $11\frac{1}{11}$

Using integer division, divide the numerator by the denominator:

$$122 \div 11 = 11r1$$

Use this result to build the equivalent mixed number, using the quotient (11) as the whole number and placing the remainder (1) over the original denominator (11):

$$11\frac{1}{11}$$

295. $\frac{12}{20}$

Begin by dividing the larger denominator by the smaller one:

$$20 \div 5 = 4$$

Now, multiply the result by the numerator:

$$4 \times 3 = 12$$

Thus:

$$\frac{3}{5} = \frac{12}{20}$$

296. $\frac{16}{56}$

Begin by dividing the larger denominator by the smaller one:

$$56 \div 7 = 8$$

Now, multiply the result by the numerator:

$$8 \times 2 = 16$$

Thus:

$$\frac{2}{7} = \frac{16}{56}$$

297. $\frac{48}{120}$

Begin by dividing the larger denominator by the smaller one:

$$120 \div 10 = 12$$

Now, multiply the result by the numerator:

$$12 \times 4 = 48$$

Thus:

$$\frac{4}{10} = \frac{48}{120}$$

298. $\frac{45}{65}$

Begin by dividing the larger denominator by the smaller one:

$$65 \div 13 = 5$$

Now, multiply the result by the numerator:

$$5 \times 9 = 45$$

Thus:

$$\frac{9}{13} = \frac{45}{65}$$

299. $\frac{72}{84}$

Begin by dividing the larger denominator by the smaller one:

$$84 \div 7 = 12$$

Now, multiply the result by the numerator:

$$12 \times 6 = 72$$

Thus:

$$\frac{6}{7} = \frac{72}{84}$$

300. $\frac{117}{135}$

Begin by dividing the larger denominator by the smaller one:

$$135 \div 15 = 9$$

Now, multiply the result by the numerator:

$$9 \times 13 = 117$$

Thus:

$$\frac{13}{15} = \frac{117}{135}$$

---

301. $\frac{4}{11}$

The numerator and denominator are both even, so you can reduce the fraction by at least a factor of 2:

$$\frac{8}{22} = \frac{4}{11}$$

---

302. $\frac{3}{7}$

The numerator and denominator are both even, so you can reduce the fraction by at least a factor of 2:

$$\frac{18}{42} = \frac{9}{21}$$

Now you can see that the resulting numerator and denominator are both divisible by 3, so you can reduce this fraction further by a factor of 3:

$$= \frac{3}{7}$$

---

303. $\frac{1}{7}$

The numerator and denominator both end in 5, so you can reduce the fraction by at least a factor of 5:

$$\frac{15}{105} = \frac{3}{21}$$

Now you can see that the resulting numerator and denominator are both divisible by 3, so you can reduce this fraction further by a factor of 3:

$$= \frac{1}{7}$$

---

304. $\frac{3}{8}$

The numerator and denominator both end in 0, so you can reduce the fraction by at least a factor of 10:

$$\frac{270}{720} = \frac{27}{72}$$

Now you can see that the resulting numerator and denominator are both divisible by 9, so you can reduce this fraction further by a factor of 9:

$$= \frac{3}{8}$$

305. $\dfrac{3}{10}$

The numerator and denominator are both divisible by 5, so you can reduce the fraction by at least a factor of 5:

$$\frac{375}{1,250} = \frac{75}{250}$$

Now you can see that the resulting numerator and denominator are still both divisible by 5, so you can reduce this fraction further by a factor of 5:

$$= \frac{15}{50}$$

Again, the resulting numerator and denominator are still both divisible by 5, so you reduce this fraction further by a factor of 5:

$$= \frac{3}{10}$$

306. $\dfrac{3}{5}$

The numerator and denominator are both even, so you can reduce the fraction by at least a factor of 2:

$$\frac{138}{230} = \frac{69}{115}$$

Now you can see that the resulting numerator is divisible by 3, but the denominator isn't (because $1 + 1 + 5 = 7$, which isn't divisible by 3). However, because 69 is divisible by 3, it has another factor that you can find by dividing:

$$69 \div 3 = 23$$

So you can rewrite the problem as follows:

$$\frac{69}{115} = \frac{3 \times 23}{115}$$

Now, check to see whether 115 is also divisible by 23:

$$115 \div 23 = 5$$

So you can rewrite the problem again as follows:

$$\frac{3 \times 23}{115} = \frac{3 \times 23}{5 \times 23}$$

Now, cancel a factor of 23 in both the numerator and denominator:

$$= \frac{3}{5}$$

307. $\dfrac{5}{7} < \dfrac{8}{11}$

Cross-multiply to check the equation:

$$\frac{5}{7} = \frac{8}{11}?$$
$$5 \times 11 = 7 \times 8?$$
$$55 = 56?$$

The resulting equation is wrong and needs to be corrected to 55 < 56, so

$$\frac{5}{7} < \frac{8}{11}$$

308. $\frac{3}{10} < \frac{4}{13}$

Cross-multiply to check the equation:

$$\frac{3}{10} = \frac{4}{13}?$$
$$3 \times 13 = 10 \times 4?$$
$$39 = 40?$$

The resulting equation is wrong and needs to be corrected to 39 < 40, so

$$\frac{3}{10} < \frac{4}{13}$$

309. $\frac{2}{5} > \frac{5}{23}$

Cross-multiply to check the equation:

$$\frac{2}{5} = \frac{5}{23}?$$
$$2 \times 23 = 5 \times 5?$$
$$46 = 25?$$

The resulting equation is wrong and needs to be corrected to 46 > 25, so

$$\frac{2}{5} > \frac{5}{23}$$

310. $\frac{5}{12} > \frac{12}{29}$

Cross-multiply to check the equation:

$$\frac{5}{12} = \frac{12}{29}?$$
$$5 \times 29 = 12 \times 12?$$
$$145 = 144?$$

The resulting equation is wrong and needs to be corrected to 145 > 144, so

$$\frac{5}{12} > \frac{12}{29}$$

311. $\frac{8}{9} = \frac{104}{117}$

Cross-multiply to check the equation:

$$\frac{8}{9} = \frac{104}{117}?$$
$$8 \times 117 = 9 \times 104$$
$$936 = 936?$$

The resulting equation is right, so

$$\frac{8}{9} = \frac{104}{117}$$

312. $\dfrac{97}{101} < \dfrac{971}{1,002}$

Cross-multiply to check the equation:

$$\frac{97}{101} = \frac{971}{1,002}?$$

$$97 \times 1,002 = 971 \times 101$$

$$97,194 = 98,071?$$

The resulting equation is wrong and needs to be corrected to 97,194 < 98,071, so

$$\frac{97}{101} < \frac{971}{1,002}$$

313. $\dfrac{12}{35}$

Multiply the two numerators to find the numerator of the answer, and multiply the two denominators to find the denominator of the answer:

$$\frac{3}{5} \times \frac{4}{7} = \frac{12}{35}$$

314. $\dfrac{1}{12}$

To begin, cancel out a factor of 2 in the numerator and denominator:

$$\frac{2}{15} \times \frac{5}{8} = \frac{1}{15} \times \frac{5}{4}$$

Next, cancel out a factor of 5:

$$= \frac{1}{3} \times \frac{1}{4}$$

Multiply the two numerators to find the numerator of the answer, and multiply the two denominators to find the denominator of the answer:

$$= \frac{1}{12}$$

315. $\dfrac{3}{40}$

To begin, cancel out a factor of 7 in the numerator and denominator:

$$\frac{7}{12} \times \frac{9}{70} = \frac{1}{12} \times \frac{9}{10}$$

Next, cancel out a factor of 3:

$$= \frac{1}{4} \times \frac{3}{10}$$

Multiply the two numerators to find the numerator of the answer, and multiply the two denominators to find the denominator of the answer:

$$= \frac{3}{40}$$

316. $\frac{135}{242}$

Multiply the two numerators to find the numerator of the answer, and multiply the two denominators to find the denominator of the answer:

$$\frac{9}{11} \times \frac{15}{22} = \frac{135}{242}$$

317. $\frac{1}{6}$

To begin, cancel out a factor of 13 in the numerator and denominator:

$$\frac{17}{39} \times \frac{13}{34} = \frac{17}{3} \times \frac{1}{34}$$

Next, cancel out a factor of 17:

$$= \frac{1}{3} \times \frac{1}{2}$$

Multiply the two numerators to find the numerator of the answer, and multiply the two denominators to find the denominator of the answer:

$$= \frac{1}{6}$$

318. $\frac{3}{10}$

Turn the division into multiplication by changing the second fraction to its reciprocal (that is, by flipping it upside-down):

$$\frac{1}{6} \div \frac{5}{9} = \frac{1}{6} \times \frac{9}{5}$$

Next, cancel out a factor of 3 in the numerator and denominator:

$$= \frac{1}{2} \times \frac{3}{5}$$

Multiply the two numerators to find the numerator of the answer, and multiply the two denominators to find the denominator of the answer:

$$= \frac{3}{10}$$

319. $3\frac{1}{16}$

Turn the division into multiplication by changing the second fraction to its reciprocal (that is, by flipping it upside-down):

$$\frac{7}{8} \div \frac{2}{7} = \frac{7}{8} \times \frac{7}{2}$$

Multiply the two numerators to find the numerator of the answer, and multiply the two denominators to find the denominator of the answer:

$$= \frac{49}{16}$$

Change the improper fraction into a mixed number by dividing ($49 \div 16 = 3r1$):

$$= 3\frac{1}{16}$$

320. $\frac{1}{16}$

Turn the division into multiplication by changing the second fraction to its reciprocal (that is, by flipping it upside-down):

$$\frac{1}{40} \div \frac{2}{5} = \frac{1}{40} \times \frac{5}{2}$$

Next, cancel out a factor of 5 in the numerator and denominator:

$$= \frac{1}{8} \times \frac{1}{2}$$

Multiply the two numerators to find the numerator of the answer, and multiply the two denominators to find the denominator of the answer:

$$= \frac{1}{16}$$

321. $1\frac{1}{2}$

Turn the division into multiplication by changing the second fraction to its reciprocal (that is, by flipping it upside-down):

$$\frac{10}{13} \div \frac{20}{39} = \frac{10}{13} \times \frac{39}{20}$$

Now, cancel out a factor of 10 in the numerator and denominator:

$$= \frac{1}{13} \times \frac{39}{2}$$

Next, cancel out a factor of 13 in the numerator and denominator:

$$= \frac{1}{1} \times \frac{3}{2}$$

Multiply the two numerators to find the numerator of the answer, and multiply the two denominators to find the denominator of the answer:

$$= \frac{3}{2}$$

Change the improper fraction into a mixed number by dividing:

$$= 1\frac{1}{2}$$

322. $1\frac{4}{5}$

Turn the division into multiplication by changing the second fraction to its reciprocal (that is, by flipping it upside-down):

$$\frac{51}{55} \div \frac{17}{33} = \frac{51}{55} \times \frac{33}{17}$$

Now, cancel out a factor of 11 in the numerator and denominator:

$$= \frac{51}{5} \times \frac{3}{17}$$

You can also cancel out a factor of 17 in the numerator and denominator:

$$= \frac{3}{5} \times \frac{3}{1}$$

Multiply the two numerators to find the numerator of the answer, and multiply the two denominators to find the denominator of the answer:

$$= \frac{9}{5}$$

Change the improper fraction into a mixed number by dividing:

$$= 1\frac{4}{5}$$

323. $1\frac{2}{7}$

When two fractions have the same denominator (bottom number), add them by adding the numerators (top numbers) and keeping the denominator the same:

$$\frac{3}{7} + \frac{6}{7} = \frac{3+6}{7} = \frac{9}{7}$$

Change the improper fraction into a mixed number:

$$= 1\frac{2}{7}$$

324. $\frac{7}{8}$

When two fractions have the same denominator (bottom number), add them by adding the numerators (top numbers) and keeping the denominator the same:

$$\frac{5}{16} + \frac{9}{16} = \frac{5+9}{16} = \frac{14}{16}$$

The numerator and denominator are both even numbers, so you can reduce this fraction by a factor of 2:

$$= \frac{7}{8}$$

325. $\frac{3}{5}$

When two fractions have the same denominator (bottom number), subtract them by subtracting the numerators (top numbers) and keeping the denominator the same:

$$\frac{34}{35} - \frac{13}{35} = \frac{34-13}{35} = \frac{21}{35}$$

The numerator and denominator are both divisible by 7, so you can reduce this fraction by a factor of 7:

$$= \frac{3}{5}$$

326. $\frac{15}{56}$

When two fractions each have a numerator (top number) of 1, you can add them using this simple trick: Add the two denominators to find the numerator of the sum, then multiply the two denominators to find the denominator of the sum:

$$\frac{1}{7} + \frac{1}{8} = \frac{7+8}{7 \times 8} = \frac{15}{56}$$

327. $\frac{1}{10}$

When two fractions each have a numerator (top number) of 1, you can subtract them using this simple trick: Subtract the larger denominator minus the smaller one to find the numerator of the difference, then multiply the two denominators to find the denominator of the difference:

$$\frac{1}{6} - \frac{1}{15} = \frac{15-6}{6 \times 15} = \frac{9}{90}$$

Because the numerator and denominator are both divisible by 9, you need to reduce this fraction by a factor of 9:

$$= \frac{1}{10}$$

328. $\frac{1}{35}$

When two fractions each have a numerator (top number) of 1, you can subtract them using this simple trick: Subtract the larger denominator minus the smaller one to find the numerator of the difference, then multiply the two denominators to find the denominator of the difference:

$$\frac{1}{10} - \frac{1}{14} = \frac{14-10}{10 \times 14} = \frac{4}{140}$$

Because the numerator and denominator are both even numbers, you need to reduce this fraction by a factor of at least 2. In this case, once you reduce by a factor of 2 you find that you can reduce by 2 again:

$$= \frac{2}{70} = \frac{1}{35}$$

329. $\frac{29}{35}$

You can add any two fractions by using cross-multiplication techniques. Cross-multiply and add the results to find the numerator of the answer, then multiply the two denominators to find the denominator of the answer:

$$\frac{2}{5} + \frac{3}{7} = \frac{14+15}{35} = \frac{29}{35}$$

330. $\frac{11}{20}$

You can subtract any two fractions by using cross-multiplication techniques. Cross-multiply and subtract the results to find the numerator of the answer, then multiply the two denominators to find the denominator of the answer:

$$\frac{3}{4} - \frac{1}{5} = \frac{15-4}{20} = \frac{11}{20}$$

331. $1\frac{11}{24}$

You can add any two fractions by using cross-multiplication techniques. Cross-multiply and add the results to find the numerator of the answer, then multiply the two denominators to find the denominator of the answer:

$$\frac{5}{8} + \frac{5}{6} = \frac{30 + 40}{48} = \frac{70}{48}$$

The numerator and denominator are both even, so you can reduce this result by a factor of 2:

$$= \frac{35}{24}$$

Now, change this improper fraction into a mixed number:

$$= 1\frac{11}{24}$$

332. $\frac{13}{18}$

You can subtract any two fractions by using cross-multiplication techniques. Cross-multiply and subtract the results to find the numerator of the answer, then multiply the two denominators to find the denominator of the answer:

$$\frac{5}{6} - \frac{1}{9} = \frac{45 - 6}{54} = \frac{39}{54}$$

The numerator and denominator are both divisible by 3 (because $3 + 9 = 12$ and $5 + 4 = 9$, both of which are divisible by 3), so you can reduce this result by a factor of 3:

$$= \frac{13}{18}$$

333. $\frac{37}{60}$

You can add any two fractions by using cross-multiplication techniques. Cross-multiply and add the results to find the numerator of the answer, then multiply the two denominators to find the denominator of the answer:

$$\frac{7}{15} + \frac{3}{20} = \frac{140 + 45}{300} = \frac{185}{300}$$

The numerator and denominator are both divisible by 5, so you can reduce this result by a factor of 5:

$$= \frac{37}{60}$$

334. $\frac{23}{187}$

You can subtract any two fractions by using cross-multiplication techniques. Cross-multiply and subtract the results to find the numerator of the answer, then multiply the two denominators to find the denominator of the answer:

$$\frac{2}{11} - \frac{1}{17} = \frac{34 - 11}{187} = \frac{23}{187}$$

335. $\dfrac{67}{100}$

You can subtract any two fractions by using cross-multiplication techniques. Cross-multiply and subtract the results to find the numerator of the answer, then multiply the two denominators to find the denominator of the answer:

$$\frac{17}{20} - \frac{9}{50} = \frac{850 - 180}{1,000} = \frac{670}{1,000}$$

The numerator and denominator are both divisible by 10, so you can reduce this result by a factor of 10:

$$= \frac{67}{100}$$

336. $1\dfrac{1}{360}$

You can add any two fractions by using cross-multiplication techniques. Cross-multiply and add the results to find the numerator of the answer, then multiply the two denominators to find the denominator of the answer:

$$\frac{1}{40} + \frac{44}{45} = \frac{45 + 1,760}{1,800} = \frac{1,805}{1,800}$$

The numerator and denominator are both divisible by 5, so you can reduce this result by a factor of 5:

$$= \frac{361}{360}$$

Now, change this improper fraction into a mixed number:

$$= 1\frac{1}{360}$$

337. $\dfrac{2}{3}$

To subtract, increase the terms of $\frac{3}{4}$ by a factor of 3:

$$\frac{3}{4} - \frac{1}{12} = \frac{9}{12} - \frac{1}{12} = \frac{8}{12}$$

Because the numerator and denominator are both even numbers, you need to reduce this fraction by a factor of at least 2. In this case, once you reduce by a factor of 2 you find that you can reduce by 2 again:

$$\frac{4}{6} = \frac{2}{3}$$

338. $1\dfrac{3}{20}$

To add, increase the terms of $\frac{4}{5}$ by a factor of 4:

$$\frac{4}{5} + \frac{7}{20} = \frac{16}{20} + \frac{7}{20} = \frac{23}{20}$$

Change the resulting improper fraction to a mixed number:

$$= 1\frac{3}{20}$$

339. $\dfrac{4}{21}$

To subtract, increase the terms of $\dfrac{3}{7}$ by a factor of 3:

$$\frac{3}{7} - \frac{5}{21} = \frac{9}{21} - \frac{5}{21} = \frac{4}{21}$$

340. $1\dfrac{3}{50}$

To add, increase the terms of $\dfrac{3}{20}$ by a factor of 5:

$$\frac{91}{100} + \frac{3}{20} = \frac{91}{100} + \frac{15}{100} = \frac{106}{100}$$

Change the resulting improper fraction to a mixed number:

$$= 1\frac{6}{100}$$

Finally, reduce the fraction:

$$= 1\frac{3}{50}$$

341. $1\dfrac{31}{117}$

To add, increase the terms of $\dfrac{12}{13}$ by a factor of 9:

$$\frac{12}{13} + \frac{40}{117} = \frac{108}{117} + \frac{40}{117} = \frac{148}{117}$$

Change the resulting improper fraction to a mixed number:

$$= 1\frac{31}{117}$$

342. $\dfrac{7}{95}$

To subtract, increase the terms of $\dfrac{35}{38}$ by a factor of 5:

$$\frac{189}{190} - \frac{35}{38} = \frac{189}{190} - \frac{175}{190} = \frac{14}{190}$$

Because the numerator and denominator are both even numbers, you need to reduce this fraction by a factor of at least 2:

$$= \frac{7}{95}$$

343. $\dfrac{11}{18}$

The lowest common denominator of 6 and 9 is 18, so increase the terms of the two fractions by 3 and 2, respectively; then subtract:

$$\frac{5}{6} - \frac{2}{9} = \frac{15}{18} - \frac{4}{18} = \frac{11}{18}$$

344. $1\frac{19}{24}$

The lowest common denominator of 8 and 12 is 24, so increase the terms of the two fractions by 3 and 2, respectively; then add:

$$\frac{7}{8} + \frac{11}{12} = \frac{21}{24} + \frac{22}{24} = \frac{43}{24}$$

Change this improper fraction to a mixed number:

$$= 1\frac{19}{24}$$

345. $1\frac{3}{50}$

The lowest common denominator of 10 and 25 is 50, so increase the terms of the two fractions by 5 and 2, respectively; then add:

$$\frac{3}{10} + \frac{19}{25} = \frac{15}{50} + \frac{38}{50} = \frac{53}{50}$$

Change this improper fraction to a mixed number:

$$= 1\frac{3}{50}$$

346. $\frac{11}{30}$

The lowest common denominator of 6 and 15 is 30, so increase the terms of the two fractions by 5 and 2, respectively:

$$\frac{5}{6} - \frac{7}{15} = \frac{25}{30} - \frac{14}{30} = \frac{11}{30}$$

347. $2\frac{1}{20}$

The lowest common denominator of 4, 5, and 10 is 20, so increase the terms of the three fractions by 5, 4, and 2, respectively:

$$\frac{3}{4} + \frac{2}{5} + \frac{9}{10} = \frac{15}{20} + \frac{8}{20} + \frac{18}{20} = \frac{41}{20}$$

Change this improper fraction into a mixed number:

$$= 2\frac{1}{20}$$

348. $\frac{49}{60}$

The lowest common denominator of 10, 12, and 15 is 60, so increase the terms of the three fractions by 6, 5, and 4, respectively:

$$\frac{7}{10} + \frac{7}{12} - \frac{7}{15} = \frac{42}{60} + \frac{35}{60} - \frac{28}{60}$$

Add the first two fractions and then subtract the third:

$$= \frac{77}{60} - \frac{28}{60} = \frac{49}{60}$$

349. $3\frac{17}{20}$

Change both mixed numbers to improper fractions; then multiply:

$$1\frac{3}{4} \times 2\frac{1}{5} = \frac{7}{4} \times \frac{11}{5} = \frac{77}{20}$$

Now, change the result back to a mixed number:

$$= 3\frac{17}{20}$$

350. $6\frac{7}{30}$

Change both mixed numbers to improper fractions; then multiply:

$$3\frac{2}{5} \times 1\frac{5}{6} = \frac{17}{5} \times \frac{11}{6} = \frac{187}{30}$$

Now, change the result back to a mixed number:

$$= 6\frac{7}{30}$$

351. $1\frac{3}{5}$

Change both mixed numbers to improper fractions:

$$6\frac{2}{3} \div 4\frac{1}{6} = \frac{20}{3} \div \frac{25}{6}$$

Now change this problem from division to multiplication by changing the second fraction to its reciprocal:

$$= \frac{20}{3} \times \frac{6}{25}$$

Cancel out common factors in the numerator and denominator:

$$= \frac{4}{3} \times \frac{6}{5} = \frac{4}{1} \times \frac{2}{5}$$

Now, multiply and change the result back to a mixed number:

$$= \frac{8}{5} = 1\frac{3}{5}$$

352. $3\frac{1}{3}$

Change both mixed numbers to improper fractions:

$$10\frac{2}{3} \div 3\frac{1}{5} = \frac{32}{3} \div \frac{16}{5}$$

Now change this problem from division to multiplication by changing the second fraction to its reciprocal:

$$= \frac{32}{3} \times \frac{5}{16}$$

Cancel out common factors in the numerator and denominator and multiply:

$$= \frac{2}{3} \times \frac{5}{1} = \frac{10}{3}$$

Change the result back to a mixed number:

$$= 3\frac{1}{3}$$

---

353. $4\frac{1}{2}$

To begin, set up the problem in column form.

Add the fractional parts and reduce the result:

$$\frac{1}{8} + \frac{3}{8} = \frac{4}{8} = \frac{1}{2}$$

Now add the whole-number part:

$$1 + 3 = 4$$

---

354. $8\frac{2}{3}$

To begin, set up the problem in column form.

Add the fractional parts and reduce the result:

$$\frac{5}{6} + \frac{5}{6} = \frac{10}{6} = \frac{5}{3}$$

Because this result is an improper fraction, change it to a mixed number:

$$= 1\frac{2}{3}$$

Carry the 1 from this mixed number into the whole-number column and add:

$$1 + 2 + 5 = 8$$

---

355. $10\frac{26}{35}$

To begin, set up the problem in column form.

Add the fractional parts using cross-multiplication:

$$\frac{3}{5} + \frac{1}{7} = \frac{21 + 5}{35} = \frac{26}{35}$$

Now add the whole-number column:

$$4 + 6 = 10$$

---

356. $10\frac{1}{12}$

To begin, set up the problem in column form.

Add the fractional parts by finding a common denominator:

$$\frac{1}{2} + \frac{5}{6} + \frac{3}{4} = \frac{6}{12} + \frac{10}{12} + \frac{9}{12} = \frac{25}{12}$$

Because this result is an improper fraction, change it to a mixed number:

$$= 2\frac{1}{12}$$

Carry the 2 from this mixed number into the whole-number column and add:

$$2 + 3 + 1 + 4 = 10$$

357. $6\frac{1}{2}$

To begin, set up the problem in column form.

Add the fractional parts by finding a common denominator (in this case, the lowest common denominator is 20):

$$\frac{1}{4} + \frac{3}{5} + \frac{7}{10} + \frac{19}{20} = \frac{5}{20} + \frac{12}{20} + \frac{14}{20} + \frac{19}{20} = \frac{50}{20}$$

Reduce the result:

$$= \frac{5}{2}$$

Because this result is an improper fraction, change it to a mixed number:

$$= 2\frac{1}{2}$$

Carry the 2 from this mixed number into the whole-number column and add:

$$2 + 1 + 1 + 1 + 1 = 6$$

358. $2\frac{3}{4}$

To begin, set up the problem in column form.

Subtract the fractional parts and reduce the result:

$$\frac{7}{8} - \frac{1}{8} = \frac{6}{8} = \frac{3}{4}$$

Subtract the whole numbers:

$$6 - 4 = 2$$

359. $3\frac{7}{18}$

To begin, set up the problem in column form.

Subtract the fractional parts and reduce the result:

$$\frac{5}{9} - \frac{1}{6} = \frac{10}{18} - \frac{3}{18} = \frac{7}{18}$$

Reduce the result:

$$\frac{7}{18}$$

Subtract the whole numbers:

$$42 - 39 = 3$$

360. $9\frac{1}{3}$

To begin, set up the problem in column form.

Subtract the fractional parts by raising the terms of the second fraction:

$$\frac{11}{15} - \frac{2}{5} = \frac{11}{15} - \frac{6}{15} = \frac{5}{15}$$

Reduce the result:

$$= \frac{1}{3}$$

Subtract the whole numbers:

$$10 - 1 = 9$$

361. $6\frac{2}{7}$

To begin, set up the problem in column form.

The first fraction $\left(\frac{1}{7}\right)$ is less than the second fraction $\left(\frac{6}{7}\right)$, so you need to borrow 1 from 9 before you can subtract.

Change the mixed number $1\frac{1}{7}$ to an improper fraction:

$$\frac{8}{7}$$

Now you can subtract the fractional parts:

$$\frac{8}{7} - \frac{6}{7} = \frac{2}{7}$$

Subtract the whole numbers:

$$8 - 2 = 6$$

362. $29\frac{81}{88}$

To begin, set up the problem in column form.

The denominators are different, so change the fractional parts of each number so that they share a common denominator:

$$\frac{33}{88} - \frac{40}{88} =$$

The first fraction $\left(\frac{33}{88}\right)$ is less than the second fraction $\left(\frac{40}{88}\right)$, so you need to borrow 1 from 78 before you can subtract.

Change the mixed number $1\frac{33}{88}$ to an improper fraction:

$$\frac{121}{88}$$

Now you can subtract the fractional parts:

$$\frac{121}{88} - \frac{40}{88} = \frac{81}{88}$$

Subtract the whole numbers:

$$77 - 48 = 29$$

---

363. $\frac{9}{10}$

Begin by solving the addition in the numerator and the subtraction in the denominator:

$$\frac{\frac{1}{5} + \frac{2}{5}}{\frac{5}{6} - \frac{1}{6}} = \frac{\frac{3}{5}}{\frac{4}{6}}$$

Reduce the fraction in the denominator:

$$= \frac{\frac{3}{5}}{\frac{2}{3}}$$

Next, rewrite this complex fraction as division of two fractions and solve:

$$\frac{3}{5} \div \frac{2}{3} = \frac{3}{5} \times \frac{3}{2} = \frac{9}{10}$$

---

364. $2\frac{5}{8}$

Begin by solving the addition in the numerator and the subtraction in the denominator:

$$\frac{1 + \frac{3}{4}}{1 - \frac{1}{3}} = \frac{\frac{7}{4}}{\frac{2}{3}}$$

Next, rewrite this complex fraction as division of two fractions and solve:

$$\frac{7}{4} \div \frac{2}{3} = \frac{7}{4} \times \frac{3}{2} = \frac{21}{8}$$

Change this improper fraction to a mixed number:

$$= 2\frac{5}{8}$$

---

365. $3\frac{3}{10}$

Begin by converting the fractions in the numerator and the denominator to common denominators (8 for the fractions in the numerator, and 12 for the fractions in the denominator):

$$\frac{2 - \frac{5}{8}}{\frac{3}{4} - \frac{1}{3}} = \frac{\frac{16}{8} - \frac{5}{8}}{\frac{9}{12} - \frac{4}{12}}$$

Now, do the subtractions in the numerator and denominator:

$$= \frac{\frac{11}{8}}{\frac{5}{12}}$$

Next, rewrite this complex fraction as division of two fractions and then turn this division into multiplication:

$$\frac{11}{8} \div \frac{5}{12} = \frac{11}{8} \times \frac{12}{5}$$

Cancel a factor of 4 in the numerator and denominator and then multiply:

$$= \frac{11}{2} \times \frac{3}{5} = \frac{33}{10}$$

Change this improper fraction to a mixed number:

$$= 3\frac{3}{10}$$

366. $\quad \frac{2}{21}$

Begin by converting the fractions in the numerator and the denominator to common denominators (63 for the fractions in the numerator, and 3 for the fractions in the denominator):

$$\frac{\frac{1}{7} + \frac{1}{9}}{3 - \frac{1}{3}} = \frac{\frac{9}{63} + \frac{7}{63}}{\frac{9}{3} - \frac{1}{3}}$$

Now, do the addition in the numerator and the subtraction in the denominator:

$$= \frac{\frac{16}{63}}{\frac{8}{3}}$$

Next, rewrite this complex fraction as division of two fractions and then turn this division into multiplication:

$$\frac{16}{63} \div \frac{8}{3} = \frac{16}{63} \times \frac{3}{8}$$

Cancel a factor of 8 in the numerator and denominator:

$$= \frac{2}{63} \times \frac{3}{1}$$

Cancel a factor of 3 in the numerator and denominator and then multiply:

$$= \frac{2}{21} \times \frac{1}{1} = \frac{2}{21}$$

367. $\quad \frac{20}{27}$

Begin by solving the subtraction in the numerator by cross-multiplication:

$$\frac{6}{7} - \frac{2}{9} = \frac{54 - 14}{63} = \frac{40}{63}$$

Next, solve the addition in the denominator by raising the terms of the first fraction by a factor of 7:

$$\frac{1}{2} + \frac{5}{14} = \frac{7}{14} + \frac{5}{14} = \frac{12}{14}$$

So, you can rewrite this complex fraction as division of two fractions:

$$\frac{\dfrac{6}{7} - \dfrac{2}{9}}{\dfrac{1}{2} + \dfrac{5}{14}} = \frac{40}{63} \div \frac{12}{14}$$

Now turn this division into multiplication:

$$= \frac{40}{63} \times \frac{14}{12}$$

Cancel a factor of 7 in the numerator and denominator:

$$= \frac{40}{9} \times \frac{2}{12}$$

Now cancel by a factor of 2:

$$= \frac{40}{9} \times \frac{1}{6}$$

You can still cancel by another factor of 2:

$$= \frac{20}{9} \times \frac{1}{3}$$

Now multiply:

$$= \frac{20}{27}$$

---

368. $1\frac{1}{8}$

Begin by solving the subtraction in the numerator by changing 13 to a fraction with 2 in the denominator:

$$13 - \frac{1}{2} = \frac{26}{2} - \frac{1}{2} = \frac{25}{2}$$

Next, solve the addition in the denominator by changing 11 to a fraction with 9 in the denominator:

$$11 + \frac{1}{9} = \frac{99}{9} + \frac{1}{9} = \frac{100}{9}$$

So, you can rewrite this complex fraction as division of two fractions:

$$\frac{13 - \dfrac{1}{2}}{11 + \dfrac{1}{9}} = \frac{25}{2} \div \frac{100}{9}$$

Change the division to multiplication:

$$= \frac{25}{2} \times \frac{9}{100}$$

Cancel a factor of 25 in the numerator and denominator and multiply:

$$= \frac{1}{2} \times \frac{9}{4} = \frac{9}{8}$$

Now change this improper fraction to a mixed number:

$$= 1\frac{1}{8}$$

369. $\frac{4}{9}$

Begin by solving the addition in the numerator:

$$1 + \frac{1}{6} + \frac{1}{9} = \frac{18}{18} + \frac{3}{18} + \frac{2}{18} = \frac{23}{18}$$

Next, solve the subtraction in the denominator:

$$3 - \frac{1}{8} = \frac{24}{8} - \frac{1}{8} = \frac{23}{8}$$

Now, rewrite this complex fraction as division of two fractions:

$$\frac{1 + \frac{1}{6} + \frac{1}{9}}{3 - \frac{1}{8}} = \frac{23}{18} \div \frac{23}{8}$$

Now turn this division into multiplication:

$$= \frac{23}{18} \times \frac{8}{23}$$

Cancel a factor of 23 in the numerator and denominator:

$$= \frac{1}{18} \times \frac{8}{1}$$

Now, cancel a factor of 2 in the numerator and denominator and then multiply:

$$= \frac{1}{9} \times \frac{4}{1} = \frac{4}{9}$$

370. $1\frac{7}{12}$

Begin by simplifying the $\dfrac{1 + \frac{2}{3}}{8}$ in the numerator:

$$\frac{1 + \frac{2}{3}}{8} = \frac{\frac{5}{3}}{8} = \frac{5}{3} \div 8 = \frac{5}{3} \times \frac{1}{8} = \frac{5}{24}$$

Now simplify the $\dfrac{5}{1 - \frac{1}{3}}$ in the denominator:

$$\frac{5}{1 - \frac{1}{3}} = \frac{5}{\frac{2}{3}} = 5 \div \frac{2}{3} = 5 \times \frac{3}{2} = \frac{15}{2}$$

Substitute these two results back into the original fraction:

$$\frac{1 - \frac{1 + \frac{2}{3}}{8}}{8 - \frac{5}{1 - \frac{1}{3}}} = \frac{1 - \frac{5}{24}}{8 - \frac{15}{2}}$$

Now, change the whole numbers in the numerator and denominator to fractions with common denominators needed for subtraction:

$$= \frac{\frac{24}{24} - \frac{5}{24}}{\frac{16}{2} - \frac{15}{2}}$$

Now, do the subtractions in both the numerator and denominator:

$$= \frac{\frac{19}{24}}{\frac{1}{2}}$$

Express this result as fraction division and then change it to multiplication:

$$\frac{19}{24} \div \frac{1}{2} = \frac{19}{24} \times \frac{2}{1}$$

Cancel by a factor of 2 in the numerator and denominator and then multiply:

$$= \frac{19}{12} \times \frac{1}{1} = \frac{19}{12}$$

Change this improper fraction to a mixed number:

$$= 1\frac{7}{12}$$

371. **See below.**

i. $\frac{1}{10}$

ii. $\frac{1}{5}$

iii. $\frac{2}{5}$

iv. $\frac{1}{2}$

v. $\frac{3}{5}$

To change a decimal to a fraction, make a fraction using the decimal as the numerator and 1 as the denominator:

i. $0.1 = \frac{0.1}{1} = \frac{1}{10}$

ii. $0.2 = \frac{0.2}{1} = \frac{2}{10} = \frac{1}{5}$

iii. $0.4 = \frac{0.4}{1} = \frac{4}{10} = \frac{2}{5}$

iv. $0.5 = \frac{0.5}{1} = \frac{5}{10} = \frac{1}{2}$

v. $0.6 = \frac{0.6}{1} = \frac{6}{10} = \frac{3}{5}$

372. See below.

i. $\dfrac{1}{100}$

ii. $\dfrac{1}{20}$

iii. $\dfrac{1}{8}$

iv. $\dfrac{1}{4}$

v. $\dfrac{3}{4}$

To change a decimal to a fraction, make a fraction using the decimal as the numerator and 1 as the denominator:

i. $0.01 = \dfrac{0.01}{1} = \dfrac{0.1}{10} = \dfrac{1}{100}$

ii. $0.05 = \dfrac{0.05}{1} = \dfrac{0.5}{10} = \dfrac{5}{100} = \dfrac{1}{20}$

iii. $0.125 = \dfrac{0.125}{1} = \dfrac{1.25}{10} = \dfrac{12.5}{100} = \dfrac{125}{1,000} = \dfrac{1}{8}$

iv. $0.25 = \dfrac{0.25}{1} = \dfrac{2.5}{10} = \dfrac{25}{100} = \dfrac{1}{4}$

v. $0.75 = \dfrac{0.75}{1} = \dfrac{7.5}{10} = \dfrac{75}{100} = \dfrac{3}{4}$

373. $\dfrac{17}{100}$

To change a decimal to a fraction, make a fraction using the decimal as the numerator and 1 as the denominator:

$$\dfrac{0.17}{1}$$

Next, multiply both the numerator and the denominator by 10 until the decimal in the numerator turns into a whole number:

$$= \dfrac{1.7}{10} = \dfrac{17}{100}$$

374. $\dfrac{7}{20}$

To change a decimal to a fraction, make a fraction using the decimal as the numerator and 1 as the denominator:

$$\dfrac{0.35}{1}$$

Next, multiply both the numerator and the denominator by 10 until the decimal in the numerator turns into a whole number:

$$= \dfrac{3.5}{10} = \dfrac{35}{100}$$

Reduce by dividing both the numerator and the denominator by 5:

$$= \dfrac{7}{20}$$

375. $\dfrac{12}{25}$

To change a decimal to a fraction, make a fraction using the decimal as the numerator and 1 as the denominator:

$$\frac{0.48}{1}$$

Next, multiply both the numerator and the denominator by 10 until the decimal in the numerator turns into a whole number:

$$= \frac{4.8}{10} = \frac{48}{100}$$

Reduce by dividing both the numerator and the denominator by 2 and then by 2 again:

$$= \frac{24}{50} = \frac{12}{25}$$

376. $\dfrac{3}{50}$

To change a decimal to a fraction, make a fraction using the decimal as the numerator and 1 as the denominator:

$$\frac{0.06}{1}$$

Next, multiply both the numerator and the denominator by 10 until the decimal in the numerator turns into a whole number:

$$= \frac{0.6}{10} = \frac{6}{100}$$

Reduce by dividing both the numerator and the denominator by 2:

$$= \frac{3}{50}$$

377. $\dfrac{87}{500}$

To change a decimal to a fraction, make a fraction using the decimal as the numerator and 1 as the denominator:

$$\frac{0.174}{1}$$

Next, multiply both the numerator and the denominator by 10 until the decimal in the numerator turns into a whole number:

$$= \frac{1.74}{10} = \frac{17.4}{100} = \frac{174}{1,000}$$

Reduce by dividing both the numerator and the denominator by 2:

$$= \frac{87}{500}$$

378. $\dfrac{1}{1,250}$

To change a decimal to a fraction, make a fraction using the decimal as the numerator and 1 as the denominator:

$$\frac{0.0008}{1}$$

Next, multiply both the numerator and the denominator by 10 until the decimal in the numerator turns into a whole number:

$$= \frac{0.008}{10} = \frac{0.08}{100} = \frac{0.8}{1,000} = \frac{8}{10,000}$$

Reduce by dividing both the numerator and the denominator by 8:

$$= \frac{1}{1,250}$$

379. $6\dfrac{7}{100}$

To change a decimal that is greater than 1 to a fraction, break off the whole-number part and make a fraction of the decimal part using the decimal as the numerator and 1 as the denominator:

$$6 + \frac{0.07}{1}$$

Next, multiply both the numerator and the denominator by 10 until the decimal in the numerator turns into a whole number:

$$= 6 + \frac{0.7}{10} = 6 + \frac{7}{100}$$

Express the result as a mixed number:

$$= 6\frac{7}{100}$$

380. $2\dfrac{101}{5,000}$

To change a decimal that is greater than 1 to a fraction, break off the whole-number part and make a fraction of the decimal part using the decimal as the numerator and 1 as the denominator:

$$2 + \frac{0.0202}{1}$$

Next, multiply both the numerator and the denominator by 10 until the decimal in the numerator turns into a whole number:

$$= 2 + \frac{0.202}{10} = 2 + \frac{2.02}{100} = 2 + \frac{20.2}{1,000} = 2 + \frac{202}{10,000}$$

Express the result as a mixed number and then reduce the fractional part:

$$= 2\frac{202}{10,000} = 2\frac{101}{5,000}$$

381. 0.13

To change this fraction to a decimal, divide the numerator and denominator by 10 until the denominator becomes 1:

$$\frac{13}{100} = \frac{1.3}{10} = \frac{0.13}{1} = 0.13$$

382. 0.0143

To change this fraction to a decimal, divide the numerator and denominator by 10 until the denominator becomes 1:

$$\frac{143}{10,000} = \frac{14.3}{1,000} = \frac{1.43}{100} = \frac{0.143}{10} = \frac{0.0143}{1} = 0.0143$$

383. 0.15

Divide $3 \div 20$, until the resulting decimal terminates:

$$
\begin{array}{r}
0.15 \\
20\overline{)3.00} \\
-20 \\
\hline
100 \\
-100 \\
\hline
0
\end{array}
$$

Therefore, $\frac{3}{20} = 0.15$.

384. 0.24

Divide $6 \div 25$, until the resulting decimal terminates:

$$
\begin{array}{r}
0.24 \\
25\overline{)6.00} \\
-50 \\
\hline
100 \\
-100 \\
\hline
0
\end{array}
$$

Therefore, $\frac{6}{25} = 0.24$.

385. 0.375

Divide $3 \div 8$, until the resulting decimal terminates:

$$
\begin{array}{r}
0.375 \\
8\overline{)3.000} \\
\underline{-24} \\
60 \\
\underline{-56} \\
40 \\
\underline{-40} \\
0
\end{array}
$$

Therefore, $\frac{3}{8} = 0.375$.

386. 0.5625

Divide $9 \div 16$, until the resulting decimal terminates:

$$
\begin{array}{r}
0.5625 \\
16\overline{)9.0000} \\
\underline{-80} \\
100 \\
\underline{-96} \\
40 \\
\underline{-32} \\
80 \\
\underline{-80} \\
0
\end{array}
$$

Therefore, $\frac{9}{16} = 0.5625$.

387. $0.\overline{3}$

Divide $1 \div 3$, until the resulting decimal repeats:

$$
\begin{array}{r}
0.33 \\
3\overline{)1.00} \\
\underline{-9} \\
10 \\
\underline{-9} \\
1
\end{array}
$$

Therefore, $\frac{1}{3} = 0.\overline{3}$.

388. $0.\overline{4}$

Divide $4 \div 9$, until the resulting decimal repeats:

$$
\begin{array}{r}
0.44 \\
9{\overline{\smash{\big)}\,4.00}} \\
\underline{-36} \\
40 \\
\underline{-36} \\
4
\end{array}
$$

Therefore, $\frac{4}{9} = 0.\overline{4}$.

389. $0.0\overline{5}$

Divide $1 \div 18$, until the resulting decimal repeats:

$$
\begin{array}{r}
0.055 \\
18{\overline{\smash{\big)}\,1.000}} \\
\underline{-90} \\
100 \\
\underline{-90} \\
10
\end{array}
$$

Therefore, $\frac{1}{18} = 0.0\overline{5}$.

390. $0.\overline{123}$

Divide $123 \div 999$, until the resulting decimal repeats:

$$
\begin{array}{r}
0.123123 \\
999{\overline{\smash{\big)}\,123.000000}} \\
\underline{-999} \\
2310 \\
\underline{-1998} \\
3120 \\
\underline{-2997} \\
1230 \\
\underline{-999} \\
2310 \\
\underline{-1998} \\
3120 \\
\underline{-2997} \\
1230
\end{array}
$$

Therefore, $\frac{123}{999} = 0.\overline{123}$.

391. $\dfrac{2}{3}$

Begin with the following equation:

$$x = 0.\overline{6}$$

Multiply both sides of the equation by 10.

$$10x = 6.\overline{6}$$

Next, subtract the first equation from the second one:

$$\begin{array}{r} 10x = 6.\overline{6} \\ -x = -0.\overline{6} \\ \hline 9x = 6 \end{array}$$

To see why this step works, remember that a repeating decimal goes on forever. So when you subtract, the repeating part of both decimals cancel each other out.

At this point, just divide both sides of the equation by 9 and simplify:

$$\frac{9x}{9} = \frac{6}{9}$$

$$x = \frac{2}{3}$$

392. $\dfrac{7}{9}$

Begin with the following equation:

$$x = 0.\overline{7}$$

Multiply both sides of the equation by 10.

$$10x = 7.\overline{7}$$

Next, subtract the first equation from the second one:

$$\begin{array}{r} 10x = 7.\overline{7} \\ -x = -0.\overline{7} \\ \hline 9x = 7 \end{array}$$

To see why this step works, remember that a repeating decimal goes on forever. So when you subtract, the repeating parts of both decimals cancel each other out.

At this point, just divide both sides of the equation by 9:

$$\frac{9x}{9} = \frac{7}{9}$$

$$x = \frac{7}{9}$$

393. $\dfrac{9}{11}$

Begin with the following equation:

$$x = 0.\overline{81}$$

Multiply both sides of the equation by 100:

$$100x = 81.\overline{81}$$

Next, subtract the first equation from the second one:

$$100x = 81.\overline{81}$$
$$-x = -0.\overline{81}$$
$$99x = 81$$

To see why this step works, remember that a repeating decimal goes on forever. So when you subtract, the repeating parts of both decimals cancel each other out.

At this point, just divide both sides of the equation by 99 and simplify:

$$\frac{99x}{99} = \frac{81}{99}$$
$$x = \frac{9}{11}$$

394. $\dfrac{497}{999}$

Begin with the following equation:

$$x = 0.\overline{497}$$

Multiply both sides of the equation by 1,000:

$$1,000x = 497.\overline{497}$$

Next, subtract the first equation from the second one:

$$1,000x = 497.\overline{497}$$
$$-x = -0.\overline{497}$$
$$999x = 497$$

To see why this step works, remember that a repeating decimal goes on forever. So when you subtract, the repeating parts of both decimals cancel each other out.

At this point, just divide both sides of the equation by 999:

$$\frac{999x}{999} = \frac{497}{999}$$
$$x = \frac{497}{999}$$

395. 4.16

Line up the decimal points and add the numbers, placing the decimal point in the same position in the answer:

$$\begin{array}{r} 3.4 \\ + \quad 0.76 \\ \hline 4.16 \end{array}$$

396. **821.739**

Line up the decimal points and add the numbers, placing the decimal point in the same position in the answer:

$$
\begin{array}{r}
821.7 \\
+\ \ 0.039 \\
\hline
821.739
\end{array}
$$

397. **69.15**

Line up the decimal points and add the numbers, placing the decimal point in the same position in the answer:

$$
\begin{array}{r}
2.35 \\
66.1 \\
+\ \ 0.7 \\
\hline
69.15
\end{array}
$$

398. **1,004.598**

Line up the decimal points and add the numbers, placing the decimal point in the same position in the answer:

$$
\begin{array}{r}
912.4 \\
60.278 \\
+\ \ 031.92 \\
\hline
1004.598
\end{array}
$$

399. **86.72**

Line up the decimal points and add the numbers, placing the decimal point in the same position in the answer:

$$
\begin{array}{r}
81.222 \\
5.4 \\
+\ \ 0.098 \\
\hline
86.720
\end{array}
$$

400. **754.74075**

Line up the decimal points and add the numbers, placing the decimal point in the same position in the answer:

$$
\begin{array}{r}
745.21 \\
8.88 \\
0.6478 \\
+\ \ 0.00295 \\
\hline
754.74075
\end{array}
$$

401. **62,661.111673**

Line up the decimal points and add the numbers, placing the decimal point in the same position in the answer:

$$
\begin{array}{r}
0.982 \\
0.009673 \\
58,433.2 \\
3,381 \\
+ \quad 845.92 \\
\hline
62,661.111673
\end{array}
$$

402. **25.2**

Line up the decimal points and subtract the numbers, placing the decimal point in the same position in the answer:

$$
\begin{array}{r}
76.5 \\
-51.3 \\
\hline
25.2
\end{array}
$$

403. **4.211**

Line up the decimal points and subtract the numbers, placing the decimal point in the same position in the answer:

$$
\begin{array}{r}
4.831 \\
-0.62 \\
\hline
4.211
\end{array}
$$

404. **2.927**

Line up the decimal points and subtract the numbers, placing the decimal point in the same position in the answer:

$$
\begin{array}{r}
7.007 \\
-4.08 \\
\hline
2.927
\end{array}
$$

405. **574.57**

Line up the decimal points and subtract the numbers, placing the decimal point in the same position in the answer. If you like, attach a trailing zero to the larger number (574.80) to make the subtraction clearer.

$$
\begin{array}{r}
574.80 \\
- \quad 0.23 \\
\hline
574.57
\end{array}
$$

406.

608.81

Line up the decimal points and subtract the numbers, placing the decimal point in the same position in the answer. If you like, attach two trailing zeros to the larger number (611.00) to make the subtraction clearer.

$$\begin{array}{r} 611.00 \\ -2.19 \\ \hline 608.81 \end{array}$$

407.

99.124

Line up the decimal points and subtract the numbers, placing the decimal point in the same position in the answer. If you like, attach three trailing zeros to the larger number (100.000) to make the subtraction clearer.

$$\begin{array}{r} 100.000 \\ -0.876 \\ \hline 99.124 \end{array}$$

408.

19,157.0064

Line up the decimal points and subtract the numbers, placing the decimal point in the same position in the answer. If you like, attach a trailing zero to the larger number (20,304.0070) to make the subtraction clearer.

$$\begin{array}{r} 20,304.0070 \\ -1,147.0006 \\ \hline 19,157.0064 \end{array}$$

409.

4.605

Multiply the two numbers as usual, disregarding the decimal points:

$$\begin{array}{r} 9.21 \\ \times0.5 \\ \hline 4605 \end{array}$$

Now, count the number of decimal places in both factors and add these together: 9.21 has 2 decimal places; 0.5 has 1 decimal place; 2 + 1 = 3. So, place the decimal point in the product to create a decimal with 3 decimal places:

$$\begin{array}{r} 9.21 \\ \times0.5 \\ \hline 4.605 \end{array}$$

Therefore, $9.21 \times 0.5 = 4.605$.

410. **1.1016**

Multiply the two numbers as usual, disregarding the decimal points:

$$
\begin{array}{r}
13.77 \\
\times\ \ 0.08 \\
\hline
11016
\end{array}
$$

Now, count the number of decimal places in both factors and add these together: 13.77 has 2 decimal places; 0.08 has 2 decimal places; 2 + 2 = 4. So, place the decimal point in the product to create a decimal with 4 decimal places:

$$
\begin{array}{r}
13.77 \\
\times\ \ 0.08 \\
\hline
1.1016
\end{array}
$$

Therefore, $13.77 \times 0.08 = 1.1016$.

411. **0.67528**

Multiply the two numbers as usual, disregarding the decimal points:

$$
\begin{array}{r}
0.0734 \\
\times\ \ 9.2 \\
\hline
1468 \\
+6606\ \ \\
\hline
67528
\end{array}
$$

Now, count the number of decimal places in both factors and add these together: 0.0734 has 4 decimal places; 9.2 has 1 decimal place; 4 + 1 = 5. So, place the decimal point in the product to create a decimal with 5 decimal places:

$$
\begin{array}{r}
0.0734 \\
\times\ \ 9.2 \\
\hline
1468 \\
+6606\ \ \\
\hline
.67528
\end{array}
$$

Therefore, $0.0734 \times 9.2 = 0.67528$.

412. **5.56686**

Multiply the two numbers as usual, disregarding the decimal points:

$$
\begin{array}{r}
1.098 \\
\times\ \ 5.07 \\
\hline
7868 \\
+5490\ \ \\
\hline
556686
\end{array}
$$

Now, count the number of decimal places in both factors and add these together: 1.098 has 3 decimal places; 5.07 has 2 decimal places; 3 + 2 = 5. So, place the decimal point in the product to create a decimal with 5 decimal places:

$$
\begin{array}{r}
1.098 \\
\times \ 5.07 \\
\hline
7868 \\
+5490 \\
\hline
5.56686
\end{array}
$$

Therefore, $1.098 \times 5.07 = 5.56686$.

413.   78.1728

Multiply the two numbers as usual, disregarding the decimal points:

$$
\begin{array}{r}
287.4 \\
\times 0.272 \\
\hline
5748 \\
20118 \\
+5748 \\
\hline
781728
\end{array}
$$

Now, count the number of decimal places in both factors and add these together: 287.4 has 1 decimal place; 0.272 has 3 decimal places; 1 + 3 = 4. So, place the decimal point in the product to create a decimal with 4 decimal places:

$$
\begin{array}{r}
287.4 \\
\times 0.272 \\
\hline
5748 \\
20118 \\
+5748 \\
\hline
78.1728
\end{array}
$$

Therefore, $287.4 \times 0.272 = 78.1728$.

414.   0.01200914

Multiply the two numbers as usual, disregarding the decimal points:

$$
\begin{array}{r}
0.014365 \\
\times 0.836 \\
\hline
86190 \\
43095 \\
+114920 \\
\hline
12009140
\end{array}
$$

Now, count the number of decimal places in both factors and add these together: 0.014365 has 6 decimal places; 0.836 has 3 decimal places; 6 + 3 = 9. So, place the decimal point in the product to create a decimal with 9 decimal places:

$$
\begin{array}{r}
0.014365 \\
\times\,0.836 \\
\hline
86190 \\
43095 \\
+114920 \\
\hline
.012009140
\end{array}
$$

Therefore, $0.014365 \times 0.836 = 0.01200914$.

---

415.   **7.2**

Begin by setting up the division as usual:

$$0.6\overline{)4.32}$$

Now, move the decimal point in the divisor (0.6) one place to the right, so the divisor becomes a whole number. At the same time, move the decimal point in the dividend (4.32) one place to the right. Then, place another decimal point just above the decimal point in the dividend:

$$6\overline{)43.2}$$

Now, divide as usual, making sure to line up the answer with the decimal point:

$$
\begin{array}{r}
7.2 \\
6\overline{)43.2} \\
-42 \\
\hline
12 \\
-12 \\
\hline
0
\end{array}
$$

Therefore, $4.32 \div 0.6 = 7.2$.

---

416.   **37.5**

Begin by setting up the division as usual:

$$0.008\overline{)0.3}$$

Now, move the decimal point in the divisor (0.008) three places to the right, so the divisor becomes a whole number. At the same time, move the decimal point in the dividend (0.3) three places to the right. Then, place another decimal point just above the decimal point in the dividend:

$$8\overline{)300.0}$$

Now, divide as usual, making sure to line up the answer with the decimal point:

$$
\begin{array}{r}
37.5 \\
8\overline{)300.0} \\
\underline{-24} \\
60 \\
\underline{-56} \\
40 \\
\underline{-40} \\
0
\end{array}
$$

Therefore, $0.3 \div 0.008 = 37.5$.

---

417.   **6,480**

Begin by setting up the division as usual:

$$0.021\overline{)136.08}$$

Now, move the decimal point in the divisor (0.021) three places to the right, so the divisor becomes a whole number. At the same time, move the decimal point in the dividend (136.08) three places to the right. Then, place another decimal point just above the decimal point in the dividend:

$$21\overline{)136080.}$$

Now, divide as usual, making sure to line up the answer with the decimal point:

$$
\begin{array}{r}
6480. \\
21\overline{)136080.} \\
\underline{-126} \\
100 \\
\underline{-84} \\
168 \\
\underline{-168} \\
0
\end{array}
$$

Therefore, $136.08 \div 0.021 = 6,480$.

---

418.   **0.030625**

Begin by setting up the division as usual:

$$1.6\overline{)0.049}$$

Now, move the decimal point in the divisor (1.6) one place to the right, so the divisor becomes a whole number. At the same time, move the decimal point in the dividend (0.049) one place to the right. Then, place another decimal point just above the decimal point in the dividend:

$$16\overline{)0.49}$$

Answers
401–500

Now, divide as usual, making sure to line up the answer with the decimal point:

$$\begin{array}{r} .03 \\ 16\overline{)0.49} \\ \underline{-48} \\ 1 \end{array}$$

Add enough zeroes to the end of the dividend so that you can continue to divide until the decimal either terminates or repeats:

$$\begin{array}{r} .030625 \\ 16\overline{)0.490000} \\ \underline{-48} \\ 100 \\ \underline{-96} \\ 40 \\ \underline{-32} \\ 80 \\ \underline{-80} \\ 0 \end{array}$$

Therefore, $0.049 \div 1.6 = 0.030625$.

---

419.   **$0.02\overline{03}$**

Begin by setting up the division as usual:

$$3.3\overline{)0.067}$$

Now, move the decimal point in the divisor (3.3) one place to the right, so the divisor becomes a whole number. At the same time, move the decimal point in the dividend (0.067) one place to the right. Then, place another decimal point just above the decimal point in the dividend:

$$33\overline{)0.67}$$

Now, divide as usual, making sure to line up the answer with the decimal. Add enough zeroes to the end of the dividend so that you can continue to divide until the decimal either terminates or repeats:

$$\begin{array}{r} .020303 \\ 33\overline{)0.670000} \\ \underline{-66} \\ 100 \\ \underline{-99} \\ 100 \\ \underline{-99} \\ 1 \end{array}$$

At this point, the decimal has begun to repeat the digits 03. Therefore, $0.067 \div 3.3 = 0.02\overline{03}$.

420. $0.0\overline{142857}$

Begin by setting up the division as usual:

$0.007\overline{)0.0001}$

Now, move the decimal point in the divisor (0.007) three places to the right, so the divisor becomes a whole number. At the same time, move the decimal point in the dividend (0.0001) three places to the right. Then, place another decimal point just above the decimal point in the dividend:

$7\overline{)0.1}$

Now, divide as usual, making sure to line up the answer with the decimal. Add enough zeroes to the end of the dividend so that you can continue to divide until the decimal either terminates or repeats:

```
      .01428571
7)0.10000000
   −7
   30
  −28
   20
  −14
   60
  −56
   40
  −35
   50
  −49
   10
   −7
    3
```

At this point, the decimal has begun to repeat the digits 142857. Therefore, $0.0001 \div 0.007 = 0.0\overline{142857}$.

421. See below.

i. 0.01

ii. 0.05

iii. 0.1

iv. 0.5

v. 1.0

To change a percent to a decimal, move the decimal point two places to the left and drop the percent sign.

422. See below.

i. 200%

ii. 20%

iii. 2%

iv. 25%

v. 75%

To change a decimal to a percent, move the decimal point two places to the right and attach a percent sign.

423. See below.

i. $\frac{1}{10}$

ii. $\frac{1}{5}$

iii. $\frac{3}{10}$

iv. $\frac{2}{5}$

v. $\frac{1}{2}$

Simple conversions from commonly used percents to fractions should be memorized. To calculate, make a fraction with the number of the percent as the numerator and 100 as the denominator, then reduce:

i. $10\% = \frac{10}{100} = \frac{1}{10}$

ii. $20\% = \frac{20}{100} = \frac{1}{5}$

iii. $30\% = \frac{30}{100} = \frac{3}{10}$

iv. $40\% = \frac{40}{100} = \frac{2}{5}$

v. $50\% = \frac{50}{100} = \frac{1}{2}$

424. See below.

i. 50%

ii. $33\frac{1}{3}\%$

iii. $66\frac{2}{3}\%$

iv. 25%

v. 75%

Simple conversions from commonly used fractions to percents should be memorized. To calculate, increase the terms of the fraction so that the denominator becomes 100, then use the numerator with a percent sign:

i. $\frac{1}{2} \times \frac{50}{50} = \frac{50}{100} = 50\%$

ii. $\frac{1}{3} \times \frac{33\frac{1}{3}}{33\frac{1}{3}} = \frac{33\frac{1}{3}}{100} = 33\frac{1}{3}\%$

iii. $\frac{2}{3} \times \frac{33\frac{1}{3}}{33\frac{1}{3}} = \frac{66\frac{2}{3}}{100} = 66\frac{2}{3}\%$

iv. $\frac{1}{4} \times \frac{25}{25} = \frac{25}{100} = 25\%$

v. $\frac{3}{4} \times \frac{25}{25} = \frac{75}{100} = 75\%$

425. **0.37**

To change a percent to a decimal, move the decimal point two places to the left and drop the percent sign.

$37\% = 0.37$

426. **1.23**

To change a percent to a decimal, move the decimal point two places to the left and drop the percent sign.

$123\% = 1.23$

427. **0.0008**

To change a percent to a decimal, move the decimal point two places to the left and drop the percent sign.

$0.08\% = 0.0008$

428. **77%**

To change a decimal to a percent, move the decimal point two places to the right and attach a percent sign.

$0.77 = 77\%$

429. **550%**

To change a decimal to a percent, move the decimal point two places to the right and attach a percent sign.

$5.5 = 550\%$

430. 0.1%

To change a decimal to a percent, move the decimal point two places to the right and attach a percent sign.

$$0.001 = 0.1\%$$

431. $\dfrac{11}{100}$

Make a fraction by placing 11 in the numerator and 100 in the denominator.

$$\dfrac{11}{100}$$

432. $\dfrac{13}{20}$

Make a fraction by placing 65 in the numerator and 100 in the denominator, and then reduce.

$$\dfrac{65}{100} = \dfrac{13}{20}$$

433. $\dfrac{11}{25}$

Make a fraction by placing 44 in the numerator and 100 in the denominator, and then reduce.

$$\dfrac{44}{100} = \dfrac{22}{50} = \dfrac{11}{25}$$

434. $\dfrac{37}{200}$

Make a fraction by placing 18.5 in the numerator and 100 in the denominator; then multiply both the numerator and denominator by 10 to remove the decimal.

$$\dfrac{18.5}{100} = \dfrac{185}{1,000}$$

Then reduce to lowest terms.

$$= \dfrac{37}{200}$$

435. $6\frac{1}{2}$

Make a fraction by placing 650 in the numerator and 100 in the denominator, and then reduce to lowest terms.

$$\dfrac{650}{100} = \dfrac{65}{10} = \dfrac{13}{2}$$

Convert this improper fraction to a mixed number.

$$= 6\frac{1}{2}$$

436. $\dfrac{3}{1,000}$

Make a fraction by placing 0.3 in the numerator and 100 in the denominator. Then multiply both the numerator and denominator by 10 to remove the decimal.

$$\frac{0.3}{100} = \frac{3}{1,000}$$

This fraction cannot be reduced further.

437. $1\dfrac{1}{8}$

Make a fraction by placing 112.5 in the numerator and 100 in the denominator. Then multiply both the numerator and denominator by 10 to remove the decimal.

$$\frac{112.5}{100} = \frac{1,125}{1,000}$$

Next, reduce to lowest terms.

$$= \frac{225}{200} = \frac{45}{40} = \frac{9}{8}$$

Convert this improper fraction to a mixed number.

$$= 1\frac{1}{8}$$

438. $\dfrac{5}{6}$

Make a fraction by placing $83\dfrac{1}{3}$ in the numerator and 100 in the denominator. Then multiply both the numerator and denominator by 3 to remove the fraction in the numerator.

$$\frac{83\frac{1}{3}}{100} = \frac{250}{300}$$

Reduce to lowest terms.

$$= \frac{25}{30} = \frac{5}{6}$$

439. 78%

Because the denominator 50 is a factor of 100, you can increase the terms of the fraction so that the denominator is 100. To do this, multiply both the numerator and denominator by 2.

$$\frac{39}{50} \times \frac{2}{2} = \frac{78}{100}$$

Now, convert to a percent.

$$= 78\%$$

440.   85%

Because the denominator 20 is a factor of 100, you can increase the terms of the fraction so that the denominator is 100. To do this, multiply both the numerator and denominator by 5.

$$\frac{17}{20} \times \frac{5}{5} = \frac{85}{100}$$

Now, convert to a percent.

$$= 85\%$$

441.   37.5%

First, to convert $\frac{3}{8}$ to a decimal, divide 3 by 8.

$$3 \div 8 = 0.375$$

Now, change this decimal to a percent by moving the decimal point two places to the right and attaching a percent sign (%).

$$= 37.5\%$$

442.   97.5%

First, to convert $\frac{39}{40}$ to a decimal, divide 39 by 40.

$$39 \div 40 = 0.975$$

Now, change this decimal to a percent by moving the decimal point two places to the right and attaching a percent sign (%).

$$= 97.5\%$$

443. $8\frac{1}{3}\%$

First, to convert $\frac{1}{12}$ to a decimal, divide 1 by 12 until you get a repeating decimal.

$$1 \div 12 = 0.08\overline{33}$$

Now, change this repeating decimal to a percent by moving the decimal point two places to the right and attaching a percent sign (%).

$$= 8.\overline{33}\%$$

Change the repeating portion of this decimal to a fraction, recalling that $0.\overline{3} = \frac{1}{3}$.

$$= 8\frac{1}{3}\%$$

444.

$45\frac{5}{11}\%$

First, to convert $\frac{5}{11}$ to a decimal, divide 5 by 11 until you get a repeating decimal.

$$5 \div 11 = 0.\overline{45}$$

Now, change this repeating decimal to a percent by moving the decimal point two places to the right and attaching a percent sign (%).

$$= 45.\overline{45}\%$$

Change the repeating portion of this decimal to a fraction and then reduce.

$$= 45\frac{45}{99}\% = 45\frac{5}{11}\%$$

445.

260%

First, change $2\frac{3}{5}$ to a mixed number.

$$2\frac{3}{5} = \frac{13}{5}$$

To convert $\frac{13}{5}$ to a decimal, divide 13 by 5.

$$13 \div 5 = 2.6$$

Now, change this decimal to a percent by moving the decimal point two places to the right and attaching a percent sign (%).

$$= 260\%$$

446.

7.77%

First, to convert $\frac{777}{10,000}$ to a decimal, divide 777 by 10,000. This is equivalent to moving the decimal point 4 places to the left.

$$777 \div 10,000 = 0.0777$$

Now, change this decimal to a percent by moving the decimal point two places to the right and attaching a percent sign (%).

$$= 7.77\%$$

447.

100.1%

First, change $1\frac{1}{1,000}$ to a mixed number.

$$1\frac{1}{1,000} = \frac{1,001}{1,000}$$

To convert $\frac{1,001}{1,000}$ to a decimal, divide 1,001 by 1,000. This is equivalent to moving the decimal point 3 places to the left.

$$1,001 \div 1,000 = 1.001$$

Now, change this decimal to a percent by moving the decimal point two places to the right and attaching a percent sign (%).

$$= 100.1\%$$

448. 10

50% equals $\frac{1}{2}$, so divide 20 by 2.

$$20 \div 2 = 10$$

449. 15

25% equals $\frac{1}{4}$, so divide 60 by 4.

$$60 \div 4 = 15$$

450. 40

20% equals $\frac{1}{5}$, so divide 200 by 5.

$$200 \div 5 = 40$$

451. 13

10% equals $\frac{1}{10}$, so divide 130 by 10.

$$130 \div 10 = 13$$

452. 33

$33\frac{1}{3}\%$ equals $\frac{1}{3}$, so divide 99 by 3.

$$99 \div 3 = 33$$

453. 24

1% equals $\frac{1}{100}$, so divide 2,400 by 100.

$$2,400 \div 100 = 24$$

454. 9

18% of 50 equals 50% of 18, which is much easier to evaluate. 50% equals $\frac{1}{2}$, so divide 18 by 2.

$$18 \div 2 = 9$$

455. 8

32% of 25 equals 25% of 32, which is much easier to evaluate. 25% equals $\frac{1}{4}$, so divide 32 by 4.

$$32 \div 4 = 8$$

456. 4

12% of $33\frac{1}{3}$% equals $33\frac{1}{3}$% of 12, which is much easier to evaluate. $33\frac{1}{3}$% equals $\frac{1}{3}$, so divide 12 by 3.

$$12 \div 3 = 4$$

457. 3.44

Convert 8% to the decimal 0.08 and then multiply by 43.

$$0.08 \times 43 = 3.44$$

458. 6.97

Convert 41% to the decimal 0.41 and then multiply by 17.

$$0.41 \times 17 = 6.97$$

459. 6.88

Convert 215% to the decimal 2.15 and then multiply by 3.2.

$$2.15 \times 3.2 = 6.88$$

460. 0.81

Convert 7.5% to the decimal 0.075 and then multiply by 10.8.

$$0.075 \times 10.8 = 0.81$$

461. 75

Turn the problem into an equation:

$$x\% \cdot 40 = 30$$

Substitute multiplication by 0.01 for %.

$$x \cdot 0.01 \cdot 40 = 30$$

Simplify by multiplying 0.01 by 40.

$$x \cdot 0.40 = 30$$

Now, divide both sides by 0.40.

$$\frac{x \cdot 0.40}{0.40} = \frac{30}{0.40}$$
$$x = 75$$

Therefore, 75% of 40 is 30.

462.   12.5

Turn the problem into an equation.

$$20 = x\% \cdot 160$$

Substitute multiplication by 0.01 for %.

$$20 = x \cdot 0.01 \cdot 160$$

Simplify by multiplying 0.01 by 160.

$$20 = x \cdot 1.60$$

Now, divide both sides by 1.60.

$$\frac{20}{1.60} = \frac{x \cdot 1.60}{1.60}$$
$$12.5 = x$$

Therefore, 20 is 12.5% of 160.

463.   288

Turn the problem into an equation.

$$72 = 25\% \cdot x$$

Change 25% into the decimal 0.25.

$$72 = 0.25 \cdot x$$

Now, divide both sides by 0.25.

$$\frac{72}{0.25} = \frac{0.25 \cdot x}{0.25}$$
$$288 = x$$

Therefore, 72 is 25% of 288.

464.   300

Turn the problem into an equation.

$$85\% \cdot x = 255$$

Change 85% into the decimal 0.85.

$$0.85 \cdot x = 255$$

Now, divide both sides by 0.85.

$$\frac{0.85 \cdot x}{0.85} = \frac{255}{0.85}$$
$$x = 300$$

Therefore, 85% of 300 is 255.

---

465. 8,500

Turn the problem into an equation.

$$71\% \cdot x = 6,035$$

Change 71% into the decimal 0.71.

$$0.71 \cdot x = 6,035$$

Now, divide both sides by 1.08.

$$\frac{0.71 \cdot x}{0.71} = \frac{6,035}{0.71}$$
$$x = 8,500$$

Therefore, 71% of 6,035 is 8,500.

---

466. 16,300

Turn the problem into an equation.

$$108\% \cdot x = 17,604$$

Change 108% into the decimal 1.08.

$$1.08 \cdot x = 17,604$$

Now, divide both sides by 1.08.

$$\frac{1.08 \cdot x}{1.08} = \frac{17,604}{1.08}$$
$$x = 16,300$$

Therefore, 108% of 16,300 is 17,604.

---

467. 96

Turn the problem into an equation:

$$x\% \cdot 2.5 = 2.4$$

Change the % into multiplication by 0.01:

$$x \cdot 0.01 \cdot 2.5 = 2.4$$

Next, multiply the two decimals on the left side of the equation:

$$x \cdot 0.025 = 2.4$$

Now, divide both sides of the equation by 0.025:

$$\frac{x \cdot 0.025}{0.025} = \frac{2.4}{0.025}$$

Therefore, 2.5 is 96% of 2.4.

---

468. **1,000**

Turn the problem into an equation.

$$99.5 = 9.95\% \cdot x$$

Change 9.95% into the decimal 0.0995.

$$99.5 = 0.0995 \cdot x$$

Now, divide both sides by 0.0995.

$$\frac{99.5}{0.0995} = \frac{0.0995 \cdot x}{0.0995}$$
$$1,000 = x$$

Therefore, 99.5 is 9.95% of 1,000.

---

469. **150**

Turn the problem into an equation:

$$\frac{1}{2} = x\% \cdot \frac{1}{3}$$

Change the % into multiplication by $\frac{1}{100}$:

$$\frac{1}{2} = x \cdot \frac{1}{100} \cdot \frac{1}{3}$$

Next, multiply the two fractions on the right side of the equation:

$$\frac{1}{2} = x \cdot \frac{1}{300}$$

Now, multiply both sides of the equation by 300:

$$\frac{1}{2} \cdot 300 = x \cdot \frac{1}{300} \cdot 300$$
$$150 = x$$

Therefore, $\frac{1}{2}$ is 150% of $\frac{1}{3}$.

---

470. **$44\frac{4}{9}$**

Turn the problem into an equation.

$$33\frac{1}{3} = 75\% \cdot x$$

Change the mixed number $33\frac{1}{3}$ into the improper fraction $\frac{100}{3}$, and change 75% into the fraction $\frac{3}{4}$.

$$\frac{100}{3} = \frac{3}{4} \cdot x$$

Now, multiply both sides by $\frac{4}{3}$.

$$\frac{100}{3} \cdot \frac{4}{3} = \frac{4}{3} \cdot \frac{3}{4} \cdot x$$

Simplify.

$$\frac{400}{9} = x$$

Now, change the improper fraction to a mixed number.

$$44\frac{4}{9} = x$$

Therefore, $33\frac{1}{3}$ is 75% of $44\frac{4}{9}$.

---

471. 2:3

Make a fraction of dogs to cats and then reduce it to lowest terms, as follows:

$$\frac{dogs}{cats} = \frac{4}{6} = \frac{2}{3}$$

The fraction $\frac{2}{3}$ is equivalent to the ratio 2:3.

---

472. 4 to 5

Make a fraction of boys to girls and then reduce it to lowest terms, as follows:

$$\frac{boys}{girls} = \frac{12}{15} = \frac{4}{5}$$

The fraction $\frac{4}{5}$ is equivalent to the ratio 4 to 5.

---

473. 7:5

Make a fraction of married people to single people and then reduce it to lowest terms, as follows:

$$\frac{married}{single} = \frac{42}{30} = \frac{7}{5}$$

The fraction $\frac{7}{5}$ is equivalent to the ratio 7:5.

---

474. 16 to 21

Make a fraction of Karina's earnings to Tamara's and then reduce it to lowest terms, as follows:

$$\frac{Karina}{Tamara} = \frac{\$32{,}000}{\$42{,}000} = \frac{32}{42} = \frac{16}{21}$$

The fraction $\frac{16}{21}$ is equivalent to the ratio 16 to 21.

475.   **7 to 11**

Make a fraction of the distances from both yesterday and today, increasing the terms of the fraction so both the numerator and denominator are integers, as follows:

$$\frac{yesterday}{today} = \frac{4.9}{7.7} = \frac{49}{77}$$

Now reduce the fraction to lowest terms.

$$= \frac{7}{11}$$

The fraction $\frac{7}{11}$ is equivalent to the ratio 7 to 11.

476.   **3:5**

Make a complex fraction of the time fulfilled on Saturday and Sunday.

$$\frac{Saturday}{Sunday} = \frac{\frac{1}{5}}{\frac{1}{3}}$$

Now, evaluate this complex fraction as fraction division.

$$= \frac{1}{5} \div \frac{1}{3}$$

Change this to multiplication by taking the reciprocal of the second fraction.

$$= \frac{1}{5} \times \frac{3}{1} = \frac{3}{5}$$

The fraction $\frac{3}{5}$ is equivalent to the ratio 3:5.

477.   **2:7**

Make a fraction of the number of managers and the *total* number of staff.

$$\frac{managers}{total} = \frac{10}{10+25} = \frac{10}{35}$$

Now, reduce the fraction to lowest terms.

$$= \frac{2}{7}$$

The fraction $\frac{2}{7}$ is equivalent to the ratio 2:7

478.   **5:6:4**

The ratio of sophomores to juniors to seniors is 10:12:8. All three of these numbers are even, so you can divide each by 2 to reduce the ratio.

$$\frac{10}{2} : \frac{12}{2} : \frac{8}{2} = 5:6:4$$

479. 4:15

Make a fraction of the number of seniors and the *total* number of students:

$$\frac{seniors}{total} = \frac{8}{10+12+8} = \frac{8}{30}$$

Now reduce the fraction to lowest terms.

$$= \frac{4}{15}$$

The fraction $\frac{4}{15}$ is equivalent to the ratio 4:15.

480. 2 to 3

Make a fraction of the number of juniors and the *combined* number of sophomores and seniors:

$$\frac{juniors}{sophomores+seniors} = \frac{12}{10+8} = \frac{12}{18}$$

Now reduce the fraction to lowest terms.

$$= \frac{2}{3}$$

The fraction $\frac{2}{3}$ is equivalent to the ratio 2 to 3.

481. 2:4:3

If one person moves from the first floor to the second floor, the ratio of first-floor residents to second-floor residents to third-floor residents becomes 4:8:6. All three of these numbers are even, so you can divide each by 2 to deduce the ratio.

$$\frac{4}{2} : \frac{8}{2} : \frac{6}{2} = 2 : 4 : 3$$

482. 4 to 1

Originally, Ann was using 2,400 watts, but then she reduced her usage by 1,800 watts, so her usage went down to 2,400 – 1,800 = 600. Make a fraction of her usage before and after as follows:

$$\frac{before}{after} = \frac{2,400}{600} = \frac{24}{6} = \frac{4}{1}$$

This fraction is equivalent to the ratio 4 to 1.

483. 6:7

Make a fraction of the building height and the *combined* height of the building and the tower:

$$\frac{building}{building + tower} = \frac{450}{450 + 75} = \frac{450}{525}$$

Now, reduce the fraction to lowest terms.

$$= \frac{90}{105} = \frac{18}{21} = \frac{6}{7}$$

The fraction $\frac{6}{7}$ is equivalent to the ratio 6:7.

---

484. 4

Make a proportion with nonregistered voters in the numerator and registered voters in the denominator; then plug in the number of registered members:

$$\frac{nonregistered}{registered} = \frac{1}{7}$$

$$\frac{nonregistered}{28} = \frac{1}{7}$$

Now, multiply both sides of this equation by 28 to cancel out the fraction on the left side.

$$\frac{nonregistered}{28} \times 28 = \frac{1}{7} \times 28$$

$$nonregistered = 4$$

Therefore, the organization has four nonregistered members.

---

485. 18

Make a proportion with windows in the numerator and doors in the denominator; then plug in the number of doors:

$$\frac{windows}{doors} = \frac{9}{2}$$

$$\frac{windows}{4} = \frac{9}{2}$$

Now, multiply both sides of this equation by 4 to cancel out the fraction on the left side.

$$\frac{windows}{4} \times 4 = \frac{9}{2} \times 4$$

$$windows = 18$$

Therefore, the house has 18 windows.

---

486. 36

Make a proportion with purchasers in the numerator and entrants in the denominator; then plug in the number of entrants:

$$\frac{purchasers}{entrants} = \frac{3}{10}$$

$$\frac{purchasers}{120} = \frac{3}{10}$$

Now, multiply both sides of this equation by 120 to cancel out the fraction on the left side:

$$\frac{purchasers}{120} \times 120 = \frac{3}{10} \times 120$$

$$purchasers = \frac{360}{10} = 36$$

Therefore, 36 people made purchases.

487. 1,815

The diet requires a 6:4:1 ratio of protein to fat to carbohydrates. Thus, its ratio of fat to total calories is 4 to (6 + 4 + 1), which is a 4:11 ratio. Make a proportion with the total in the numerator and fat in the denominator; then plug in the number of fat calories:

$$\frac{total}{fat} = \frac{11}{4}$$

$$\frac{total}{660} = \frac{11}{4}$$

Now, multiply both sides of this equation by 660 to cancel out the fraction on the left side.

$$\frac{total}{660} \times 660 = \frac{11}{4} \times 660$$

$$total = 1,815$$

Thus, the diet permits 1,815 total calories.

488. 14

The project manager estimates that her newest project will require a 2:9 ratio of team leaders to programmers. Thus, this is a 2 to (2 + 9) ratio of team leaders to total members, which is a 2:11 ratio.

Make a proportion with the team leaders in the numerator and the total members in the denominator; then plug in the total number:

$$\frac{leaders}{total} = \frac{2}{11}$$

$$\frac{leaders}{77} = \frac{2}{11}$$

Now, multiply both sides of this equation by 77 to cancel out the fraction on the left side.

$$\frac{leaders}{77} \times 77 = \frac{2}{11} \times 77$$

$$leaders = 14$$

Thus, the project manager will need 14 team leaders.

489. **104**

Make a proportion with dinner customers in the numerator and lunch customers in the denominator; then plug in the number of lunch customers:

$$\frac{dinner}{lunch} = \frac{8}{5}$$
$$\frac{dinner}{40} = \frac{8}{5}$$

Now, multiply both sides of this equation by 40 to cancel out the fraction on the left side:

$$\frac{dinner}{40} \times 40 = \frac{8}{5} \times 40$$
$$dinner = 64$$

Thus, the diner has an average of 64 dinner customers, so the total number of customers for both lunch and dinner is 40 + 64 = 104.

490. **1,140**

Make a proportion with fiction books in the numerator and nonfiction books in the denominator; then plug in the number of nonfiction books:

$$\frac{fiction}{nonfiction} = \frac{4}{15}$$
$$\frac{fiction}{900} = \frac{4}{15}$$

Now, multiply both sides of this equation by 900 to cancel out the fraction on the left side.

$$\frac{fiction}{900} \times 900 = \frac{4}{15} \times 900$$
$$fiction = 240$$

Thus, the bookmobile has 240 fiction books, so the total number of fiction and nonfiction books is 240 + 900 = 1,140.

491. **150**

The organization has a 5:3:2 ratio of members from, respectively, Massachusetts, Vermont, and New Hampshire. Thus, it has a 5:2 ratio of members from Massachusetts to members from New Hampshire. Make a proportion with Massachusetts in the numerator and New Hampshire in the denominator; then plug in the number of members from New Hampshire:

$$\frac{M}{N} = \frac{5}{2}$$
$$\frac{M}{60} = \frac{5}{2}$$

Now, multiply both sides of this equation by 60 to cancel out the fraction on the left side.

$$\frac{M}{60} \times 60 = \frac{5}{2} \times 60$$
$$M = 150$$

Thus, the organization has 150 members from Massachusetts.

492.   98

The organization has a 5:3:2 ratio of members from, respectively, Massachusetts, Vermont, and New Hampshire. Thus, the ratio of members from Vermont to members from Massachusetts or New Hampshire is 3 to $(5 + 2)$, which is 3:7. Make a proportion with Massachusetts plus New Hampshire in the numerator and Vermont in the denominator; then plug in the number of members from Vermont:

$$\frac{M + N}{V} = \frac{7}{3}$$

$$\frac{M + N}{42} = \frac{7}{3}$$

Now, multiply both sides of this equation by 42 to cancel out the fraction on the left side.

$$\frac{M + N}{42} \times 42 = \frac{7}{3} \times 42$$

$$M + N = 98$$

Thus, the organization has 98 members from either Massachusetts or New Hampshire.

493.   72

The organization has a 5:3:2 ratio of members from, respectively, Massachusetts, Vermont, and New Hampshire. Thus, the ratio of members from Vermont to the total number of members is 3 to $(5 + 3 + 2)$, which is 3:10. Make a proportion with Vermont in the numerator and the total in the denominator; then plug in the total number of members:

$$\frac{V}{Total} = \frac{3}{10}$$

$$\frac{V}{240} = \frac{3}{10}$$

Now, multiply both sides of this equation by 240 to cancel out the fraction on the left side.

$$\frac{V}{240} \times 240 = \frac{3}{10} \times 240$$

$$V = 72$$

Thus, the organization has 72 members from Vermont.

494.   90

The ratio of Jason's laps to Anton's laps is 9 to 5, so the ratio of Jason's laps to the total laps is 9 to 14. Make a proportion with Jason in the numerator and the total laps in the denominator; then plug in 140 for the total number of laps:

$$\frac{Jason}{total} = \frac{9}{14}$$

$$\frac{Jason}{140} = \frac{9}{14}$$

Multiply both sides by 140 to get rid of the fraction on the left side:

$$\frac{Jason}{140} \times 140 = \frac{9}{14} \times 140$$

$$Jason = 90$$

Therefore, Jason swam 90 laps.

495. **$50,000**

The ratio of domestic sales to foreign sales is 6 to 1, so the ratio of foreign sales to total sales is 1 to 7.

Make a proportion with foreign sales in the numerator and total sales in the denominator; then plug in $350,000 for the total amount of revenue:

$$\frac{foreign}{total} = \frac{1}{7}$$

$$\frac{foreign}{350,000} = \frac{1}{7}$$

Now, multiply both sides of this equation by 350,000 to cancel out the fraction on the left side.

$$\frac{foreign}{350,000} \times 350,000 = \frac{1}{7} \times 350,000$$

$$foreign = 50,000$$

496. **56**

The restaurant sells a 5 to 3 ratio of red wine to white wine. So, in terms of the ratio, its total sales are 5 + 3 = 8, and the difference between its red wine sales and its white wine sales is 5 − 3 = 2. Thus, the restaurant has an 8 to 2 ratio regarding the total sales and the difference in red and white wine sales, which simplifies to a 4 to 1 ratio. Make a proportion and then fill in the difference in sales as follows:

$$\frac{total}{difference} = \frac{4}{1}$$

$$\frac{total}{14} = \frac{4}{1}$$

Now, multiply both sides of the equation by 14 to cancel out the fraction on the left side.

$$\frac{total}{14} \times 14 = \frac{4}{1} \times 14$$

$$total = 56$$

497. **50 to 53**

The portfolio began with 100% of funds and rose to 106% of value, so make a proportion of these values:

$$\frac{start}{end} = \frac{100\%}{106\%}$$

Cancel the percentages; then reduce.

$$= \frac{100}{106} = \frac{50}{53}$$

498. $60

Make a proportion of dollars to francs; then reduce:

$$\frac{dollars}{francs} = \frac{500}{450} = \frac{50}{45} = \frac{10}{9}$$

Thus, the ratio of dollars to francs in any exchange is 10:9. Now, using this ratio, make an equation and plug in 54 for the number of francs that Karl returned with.

$$\frac{dollars}{francs} = \frac{10}{9}$$

$$\frac{dollars}{54} = \frac{10}{9}$$

Multiply both sides of this equation by 54 to cancel out the fraction on the left side.

$$\frac{dollars}{54} \times 54 = \frac{10}{9} \times 54$$

$$dollars = 60$$

499. $1,000

Charles spends 20% of his income on rent and 15% on transportation, so he spends the remaining 65% on everything else. Thus, the proportion of rent to everything else is 20:65, which simplifies to 4:13.

Make a proportion of rent to everything else; then plug in 3,250 for everything else:

$$\frac{rent}{everything\ else} = \frac{4}{13}$$

$$\frac{rent}{3,250} = \frac{4}{13}$$

Now, multiply both sides of the equation by 3,250 to get rid of the fraction on the left side.

$$\frac{rent}{3,250} \times 3,250 = \frac{4}{13} \times 3,250$$

$$rent = 1,000$$

Therefore, his rent is $1,000.

500. 4

Multiplication in the alternative universe is proportional to our multiplication. In the alternative universe $\frac{1}{2} \times 3 = 2$, but in our universe, $\frac{1}{2} \times 3 = 1.5$. So, make a proportion of these two values as follows:

$$\frac{alternative}{ours} = \frac{2}{1.5}$$

Simplify this proportion by multiplying both the numerator and denominator by 2.

$$\frac{alternative}{ours} = \frac{4}{3}$$

In our universe, $\frac{1}{4} \times 12 = 3$, so plug this value into the preceding equation:

$$\frac{alternative}{3} = \frac{4}{3}$$

Now, multiply both sides of the equation by 3 to get rid of the fractions.

$$\frac{alternative}{3} \times 3 = \frac{4}{3} \times 3$$

$$alternative = 4$$

Therefore, in the alternative universe, $\frac{1}{4} \times 12 = 4$.

---

501. $\frac{7}{24}$

To find the total fraction of candy that was bought, add the two fractions. Because the two fractions both have 1 in the numerator, you can add them quickly: Add the two denominators $(8 + 6 = 14)$ to find the numerator of the answer, then multiply the two denominators $(8 \times 6 = 48)$ to find the denominator of the answer.

$$\frac{1}{8} + \frac{1}{6} = \frac{14}{48}$$

Reduce the fraction by dividing both the numerator and the denominator by 2.

$$= \frac{7}{24}$$

---

502. $\frac{1}{10}$ **mile**

To find the difference between the distances the girls ran, subtract the smaller fraction from the larger one. Subtract $\frac{3}{5}$ minus $\frac{1}{2}$ using cross-multiplication techniques:

$$\frac{3}{5} - \frac{1}{2} = \frac{6-5}{10} = \frac{1}{10}$$

---

503. $\frac{2}{15}$

The word *of* in a fraction word problem means multiplication, so multiply $\frac{1}{5}$ by $\frac{2}{3}$:

$$\frac{1}{5} \times \frac{2}{3} = \frac{2}{15}$$

---

504. $\frac{3}{20}$

To find the amount of land in each subdivision, divide the fraction by 4. To divide $\frac{3}{5}$ by 4, multiply it by its reciprocal, which is $\frac{1}{4}$:

$$\frac{3}{5} \div 4 = \frac{3}{5} \times \frac{1}{4} = \frac{3}{20}$$

Answers 501–600

505. $\frac{11}{16}$ **miles**

To find half of $1\frac{3}{8}$ miles, first convert $1\frac{3}{8}$ from a mixed number to an improper fraction:

$$1\frac{3}{8} = \frac{11}{8}$$

Now, divide by 2:

$$\frac{11}{8} \div 2 = \frac{11}{8} \times \frac{1}{2} = \frac{11}{16}$$

506. $4\frac{2}{3}$

Divide to find each child's portion of cookies. To divide 14 by 3, make an improper fraction with 14 in the numerator and 3 in the denominator; then turn it into a mixed number:

$$\frac{14}{3} = 4\frac{2}{3}$$

507. $\frac{1}{1,024}$

The word *of* in a fraction word problem means multiplication, so multiply the four fractions:

$$\frac{1}{2} \times \frac{1}{4} \times \frac{1}{8} \times \frac{1}{16} = \frac{1}{1,024}$$

508. $\frac{7}{15}$

First, calculate what part of the distance Arnold and Marion drove together:

$$\frac{1}{5} + \frac{1}{3} = \frac{8}{15}$$

Next, calculate how much farther they had to drive by subtracting this amount from 1:

$$1 - \frac{8}{15} = \frac{15}{15} - \frac{8}{15} = \frac{7}{15}$$

509. **12 hours**

Jake practices for $1\frac{1}{2}$ hours 5 days a week, and for $2\frac{1}{4}$ hours 2 times a week, so calculate as follows:

$$\left(5 \times 1\frac{1}{2}\right) + \left(2 \times 2\frac{1}{4}\right)$$

Convert both mixed numbers to improper fractions:

$$= \left(5 \times \frac{3}{2}\right) + \left(2 \times \frac{9}{4}\right)$$

Solve:

$$= \frac{15}{2} + \frac{9}{2} = \frac{24}{2} = 12$$

Therefore, Jake practices basketball for 12 hours every week.

510. 5

The pizza had 16 slices. Jeff took $\frac{1}{4}$ of these, so he took 4 slices, leaving 12. Molly took 2 more slices, leaving 10. Tracy took half of the remaining slices, so she took 5 and left 5.

511. $10\frac{1}{20}$ miles

Calculate by converting all three mixed numbers to improper fractions and then adding:

$$\frac{5}{2} + \frac{13}{4} + \frac{43}{10}$$

Change each fraction to a common denominator of 20:

$$= \frac{50}{20} + \frac{65}{20} + \frac{86}{20} = \frac{201}{20}$$

Convert the result back to a mixed number:

$$= 10\frac{1}{20}$$

512. $4\frac{3}{4}$

First, add the lengths that Esther has already found:

$$3\frac{1}{4} + 4\frac{1}{2} = 7\frac{3}{4}$$

Now, subtract this result from the amount she needs to build the shelves:

$$12\frac{1}{2} - 7\frac{3}{4} = 4\frac{3}{4}$$

Therefore, she needs an additional $12\frac{1}{2} - 7\frac{3}{4} = 4\frac{3}{4}$ feet of wood.

513. $\frac{9}{16}$ gallon

Nate drank $\frac{1}{4}$ of the gallon on Monday, so he left $\frac{3}{4}$ of the gallon. Then on Tuesday, he drank $\frac{1}{4}$ of what was left, which was:

$$\frac{1}{4} \times \frac{3}{4} = \frac{3}{16}$$

Thus, on Tuesday, he drank $\frac{3}{16}$ of a gallon from a container that held $\frac{3}{4}$ of a gallon, so he left behind:

$$\frac{3}{4} - \frac{3}{16} = \frac{12}{16} - \frac{3}{16} = \frac{9}{16}$$

Thus, he left behind $\frac{9}{16}$ of a gallon.

514. $7\frac{1}{2}$ **pounds**

First, figure out how many batches you need to make by dividing the number of cookies you need (150) by the number in each batch (25):

$$150 \div 25 = 6$$

Now, multiply the amount of butter in each batch ($1\frac{1}{4}$ pounds) by 6:

$$1\frac{1}{4} \times 6 = \frac{5}{4} \times 6 = \frac{30}{4}$$

Reduce this fraction; then change it to a mixed number:

$$= \frac{15}{2} = 7\frac{1}{2}$$

515. $\frac{3}{40}$ **gallon**

First, convert $1\frac{1}{2}$ gallons to an improper fraction ($\frac{3}{2}$ gallons); then divide it by both 5 and 4:

$$\frac{3}{2} \div 5 = \frac{3}{2} \times \frac{1}{5} = \frac{3}{10}$$

$$\frac{3}{2} \div 4 = \frac{3}{2} \times \frac{1}{4} = \frac{3}{8}$$

Next, subtract to find the difference:

$$\frac{3}{8} - \frac{3}{10} = \frac{15}{40} - \frac{12}{40} = \frac{3}{40}$$

516. $3\frac{3}{4}$ **hours**

To find how many words Harry can write in an hour, divide the number of words by the number of hours:

$$650 \div 3\frac{1}{4}$$

Calculate by changing the mixed number to an improper fraction and then changing division to multiplication:

$$650 \div \frac{13}{4} = 650 \times \frac{4}{13}$$

You can simplify this calculation by canceling a factor of 13 in both the numerator and denominator:

$$50 \times \frac{4}{1} = 200$$

Thus, Harry can write 200 words per hour. To calculate how many hours he needs to write 750 words, divide 750 by 200:

$$750 \div 200 = 3\frac{3}{4}$$

Therefore, Harry needs $3\frac{3}{4}$ hours to write a 750-word article.

517. $1\frac{7}{12}$

Craig ate $\frac{1}{4}$ of the apple pie, so he left $\frac{3}{4}$ of it. His mom ate $\frac{1}{6}$ of the blueberry pie, so she left $\frac{5}{6}$ of it. So, add the two parts that they didn't eat as follows:

$$\frac{3}{4} + \frac{5}{6} = \frac{9}{12} + \frac{10}{12} = \frac{19}{12}$$

Change this improper fraction into a mixed number:

$$= 1\frac{7}{12}$$

518. $\frac{1}{3}$

David's piece was $\frac{1}{6}$ of the cake, which left $\frac{5}{6}$ of the cake untouched. Then, Sharon cut $\frac{1}{5}$ of what was left, so calculate this amount as follows:

$$\frac{5}{6} \times \frac{1}{5} = \frac{1}{6}$$

Thus, Sharon also ate $\frac{1}{6}$ of the cake. So you can calculate what David and Sharon ate as follows:

$$\frac{1}{6} + \frac{1}{6} = \frac{2}{6} = \frac{1}{3}$$

So, David and Sharon ate $\frac{1}{3}$ of the cake, leaving $\frac{2}{3}$. Armand ate $\frac{1}{2}$ of this, so he ate $\frac{1}{3}$ of the cake and left behind $\frac{1}{3}$.

519. $7\frac{1}{2}$

An hour is 60 minutes, which is 10 times as long as 6 minutes, so multiply $\frac{3}{4}$ by 10:

$$\frac{3}{4} \times 10 = \frac{30}{4}$$

Reduce and then convert the improper fraction into a mixed number:

$$= \frac{15}{2} = 7\frac{1}{2}$$

520. 7

The trick here is to think of easier numbers and then see what happens when you double them: For example, suppose you knew that 1 chicken could lay 1 egg in 1 day. Then, if you had 2 chickens, they could lay 2 eggs in the same amount of time — that is, in 1 day.

Now, apply this same thinking to the problem: If $1\frac{1}{2}$ chickens can lay $1\frac{1}{2}$ eggs in $1\frac{1}{2}$ days, then if you had 3 chickens, they could lay 3 eggs in the same amount of time — that is, $1\frac{1}{2}$ days. Or, similarly, if you had $3\frac{1}{2}$ chickens, they could lay $3\frac{1}{2}$ eggs, again, in the same amount of time — $1\frac{1}{2}$ days.

So now, if you double the amount of time to 3 days, those same $3\frac{1}{2}$ chickens would double their output to 7 eggs.

521. **5.6 kilos**

To begin, add up the number of kilos of chocolate that Connie bought:

$$2.7 + 4.9 + 3.6 = 11.2$$

Then, divide this amount by 2:

$$11.2 \div 2 = 5.6$$

Therefore, Connie ended up with 5.6 kilos of chocolate.

522. **0.87 m**

Calculate by subtracting Blair's height, 0.97, from his father's height, 1.84:

$$1.84 - 0.97 = 0.87$$

523. **60.9 m**

Calculate by multiplying the number of meters in a step, 0.7, by the number of steps, 87:

$$0.7 \times 87 = 60.9$$

524. **82 seconds**

Divide the total number of gallons, 861, by the rate at which the water is filling the tank, 10.5:

$$861 \div 10.5 = 82$$

525. **1.3 miles**

Ed ran a total of $3.4 \times 3 = 10.2$ miles, and Heather ran a total of $2.3 \times 5 = 11.5$ miles. Calculate how much farther Heather ran by subtracting their total distances:

$$11.5 - 10.2 = 1.3$$

Therefore, Heather ran 1.3 miles farther than Ed.

526. **32.5**

To find out how many miles per gallon Myra got, divide the total number of miles she drove, 403, by the total number of gallons of gas she used, 12.4:

$$403 \div 12.4 = 32.5$$

527. **1.85**

Calculate by dividing the total number of pages, 111, by the total amount of time, 1 hour or 60 minutes:

$$111 \div 60 = 1.85$$

528.   **4.55**

Calculate by multiplying the number of liters in each can, 1.3, by the number of cans, 3.5:

$$1.3 \times 3.5 = 4.55$$

529.   **$1,824.60**

Tony paid $356.10 per month for 36 months, so he paid a total of $356.10 \times 36 = $12,819.60. Subtract the sticker price of $10,995 from this amount:

$$\$12,819.60 - \$10,995 = \$1,824.60$$

530.   **1.2 seconds**

First, calculate Ronaldo's total time by adding:

$$12.6 + 12.3 + 13.1 = 38.0$$

Next, calculate Keith's time:

$$11.8 + 12.4 + 12.6 = 36.8$$

Subtract Ronaldo's time from Keith's time:

$$38.0 - 36.8 = 1.2$$

531.   **$31.25**

First, divide $187.50 by 3 to find the cost of one day:

$$\$187.50 \div 3 = \$62.50$$

Now, divide this result by 2 to find the cost of half a day:

$$\$62.50 \div 2 = \$31.25$$

Therefore, Dora should pay $31.25.

532.   **$59.50**

Calculate the total amount that Stephanie would have paid if she had paid $6.50 for each of the 29 days she went to the pool by multiplying:

$$\$6.50 \times 29 = \$188.50$$

Find how much she saved by subtracting what she paid for the pass, $129, from the preceding result:

$$\$188.50 - 129 = \$59.50$$

Therefore, she saved $59.50.

533. $240.00

The cost for a child between 6 and 12 is $57.60 ÷ 2 = $28.80, and the cost for a child under 6 is $57.60 ÷ 3 = $19.20.

Calculate the cost for 2 adults as follows:

$57.60 × 2 = $115.20

Calculate the cost for 3 children between 6 and 12 as follows:

$28.80 × 3 = $86.40

Calculate the cost for 2 children under 6 as follows:

$19.20 × 2 = $38.40

Add up these three results:

$115.20 + $86.40 + $38.40 = $240.00

534. 37.5 mph

Secretariat ran 1.5 miles in 2 minutes and 24 seconds, which equals 144 seconds (because $2 \times 60 + 24 = 144$), so calculate how many seconds it would take him to run one mile as follows:

$$\frac{144}{1.5} = 96$$

Thus, Secretariat ran at a rate of 1 mile in 96 seconds. An hour contains 3,600 seconds (because $60 \times 60 = 3,600$), so calculate how many miles he could have run in one hour as follows:

$3,600 ÷ 96 = 37.5$

Thus, Secretariat ran the Belmont Stakes at an average rate of 37.5 miles per hour.

535. 3.05 miles

On Monday, Anita swam 0.8 miles. On Tuesday, she swam $0.8 \times 0.25 = 0.2$ miles farther than on Monday, so she swam 0.8 + 0.2 = 1 mile. On Wednesday, she swam $1 \times 0.25 = 0.25$ miles farther than on Tuesday, so she swam 1 + 0.25 = 1.25 miles. Therefore, she swam 0.8 + 1 + 1.25 = 3.05 miles.

536. 6 hours

Angela spent 15 hours in total, and 40% of this time working with her flash cards, so you want to calculate 40% of 15:

$0.4 \times 15 = 6$

Therefore, Angela spent 6 hours working with her flash cards.

537. **0.99 kilos**

Ten percent of 1.1 is 0.11 ($0.1 \times 1.1 = 0.11$), so subtract this amount from the weight of the competitor's laptop:

$$1.1 - 0.11 = 0.99$$

538. **35%**

Make a fraction of the two numbers and then reduce:

$$\frac{700}{2000} = \frac{7}{20}$$

Convert this number to a decimal by dividing; then convert to a percent:

$$7 \div 20 = 0.35 = 35\%$$

539. **20%**

Beth received a raise of $13.80 − $11.50 = $2.30. Calculate the percentage by making a fraction with $2.30 in the numerator and $11.50 in the denominator and reducing:

$$\frac{\$2.30}{\$11.50} = \frac{230}{1150} = \frac{23}{115} = \frac{1}{5}$$

This fraction equals 0.2, which equals 20%.

540. **297.5 miles**

The trip was 850 miles, and Geoff drove 35% of it the first day, so you want to calculate 35% of 850:

$$0.35 \times 850 = 297.5$$

Therefore, Geoff drove 297.5 miles the first day.

541. **231**

The book was 420 pages, and Nora read 55% of it the first day, so you want to calculate 55% of 420:

$$0.55 \times 420 = 231$$

Therefore, Nora read 231 pages.

542. **12**

Kenneth mowed the lawn 25 times, and 52% of this work was in May and June. Therefore, 48% was from July to September. You can calculate 48% of 25 easily as 25% of 48, as follows:

$$48 \div 4 = 12$$

Therefore, Kenneth mowed the lawn 12 times from July to September.

543. **19.5 minutes**

The 60-minute show has 32.5% commercials, so calculate 32.5% of 60:

$$0.325 \times 60 = 19.5$$

Therefore, the show has 19.5 minutes of commercials.

544. **20%**

Jason spent 3 hours and 45 minutes in total. Three hours is equal to 180 minutes (because $60 \times 3 = 180$), so he spent $180 + 45 = 225$ minutes altogether. He spent 45 minutes of this on the windows, so make the fraction $\frac{45}{225}$ and convert it to a percentage as follows:

$$\frac{45}{225} = \frac{9}{45} = \frac{1}{5} = 20\%$$

545. **18.75%**

Eve received a total of $8,000, of which $1,500 was from the scholarship, so make a fraction of these two numbers and reduce it as follows:

$$\frac{1,500}{8,000} = \frac{15}{80} = \frac{3}{16}$$

Now, convert this fraction to a decimal and then a percent:

$$3 \div 16 = 0.1875 = 18.75\%$$

546. **72.5%**

Janey's goal is 400 hours, of which she has completed 290. Thus, make a fraction of these two numbers and reduce it as follows:

$$\frac{290}{400} = \frac{29}{40}$$

Now, convert this fraction to a decimal and then a percent:

$$29 \div 40 = 0.725 = 72.5\%$$

547. **300 hours**

Steven studied Italian for 45 hours, which represented 15% of his preparation time. Thus, you want to solve the percent problem, "15% of what number is 45?" Turn the problem into an equation:

$$15\% \cdot x = 45$$

Change the percent to a decimal:

$$0.15 \cdot x = 45$$

Now, divide both sides by 0.15:

$$\frac{0.15 \cdot x}{0.15} = \frac{45}{0.15}$$
$$x = 300$$

Therefore, 15% of 300 hours is 45 hours.

548.

125 m

The atrium is 6.25 meters, which represents 5% of the height of the building. Thus, you want to solve the percent problem, "5% of what number is 6.25?" Turn the problem into an equation:

$$5\% \cdot x = 6.25$$

Change the percent to a decimal:

$$0.05 \cdot x = 6.25$$

Now, divide both sides by 0.05:

$$\frac{0.05 \cdot x}{0.05} = \frac{6.25}{0.05}$$
$$x = 125$$

Therefore, 5% of 125 is 6.25.

549.

$6,200

Karan's mortgage payment is $1,736, which represents 28% of her monthly income. Thus, you want to solve the percent problem, "28% of what number is 1,736?" Turn the problem into an equation:

$$28\% \cdot x = 1,736$$

Change the percent to a decimal:

$$0.28 \cdot x = 1,736$$

Now, divide both sides by 0.28:

$$\frac{0.28 \cdot x}{0.28} = \frac{1,736}{0.28}$$
$$x = 6,200$$

Therefore, 28% of $6,200 is $1,736.

550.

$60,000

Madeleine earns $135,000, which represents 225% of her previous earnings. Thus, you want to solve the percent problem, "225% of what number is 135,000?" Turn the problem into an equation:

$$225\% \cdot x = 135,000$$

Change the percent to a decimal:

$$2.25 \cdot x = 135,000$$

Now, divide both sides by 2.25:

$$\frac{2.25 \cdot x}{2.25} = \frac{135,000}{2.25}$$
$$x = 60,000$$

Therefore, 225% of $60,000 is $135,000.

551. $13,200

A percent increase of 10% is equivalent to 110% of the original amount, so you want to calculate 110% of $12,000:

$$1.1 \times \$12,000 = \$13,200$$

552. $637.50

A percent decrease of 15% is equivalent to 85% of the original amount, so you want to calculate 85% of $750:

$$0.85 \times \$750 = \$637.50$$

553. $31

A percent increase of 18% is equivalent to 118% of the original amount, so you want to calculate 118% of $26.00:

$$1.18 \times \$26 = \$30.68$$

This amount rounds up to $31.

554. $222,000

A percent decrease of 3% is equivalent to 97% of the original amount, so you want to calculate 97% of $229,000:

$$0.97 \times \$229,000 = \$222,130$$

This amount rounds down to $222,000.

555. $9.43

A percent increase of 15% is equivalent to 115% of the original amount, so you want to calculate 115% of $8.20:

$$1.15 \times \$8.20 = \$9.43$$

556. $4,866.25

A percent increase of 14.5% is equivalent to 114.5% of the original amount, so you want to calculate 114.5% of $4,250:

$$1.145 \times \$4,250 = \$4,866.25$$

557. 3.225 g

A percent increase of 7.5% is equivalent to 107.5% of the original amount, so you want to calculate 107.5% of 3:

$$1.075 \times 3 = 3.225$$

558. $17,690.40

Marian received a 9% discount on an $18,000 car, so calculate the before-tax price as 91% of $18,000:

$$0.91 \times \$18,000 = \$16,380$$

Then, 8% of this price was added on, so calculate the after-tax price as 108% of $16,380:

$$1.08 \times \$16,380 = \$17,690.40$$

559. 8%

Dane invested $7,200 and walked away with $6,624. Make a fraction of these two numbers:

$$\frac{6,642}{7,200}$$

To turn this fraction into a percent, divide the numerator by the denominator; then convert the resulting decimal to a percent:

$$6,624 \div 7,200 = 0.92 = 92\%$$

This result of 92% represents an 8% decrease from the original 100%.

560. $27.50

A percent increase of 18% is equivalent to 118% of the original amount. Thus, 118% of some number is $32.45, so set up the equation as follows:

$$118\% \cdot x = 32.45$$

Change the percent to a decimal:

$$1.18 \cdot x = 32.45$$

Now, divide both sides by 1.18:

$$\frac{1.18 \cdot x}{1.18} = \frac{32.45}{1.18}$$
$$x = 27.5$$

Therefore, 118% of $27.50 is $32.45.

561. 1.776×10^3

Begin by multiplying 1,776 by 10^0 (recall that $10^0 = 1$, so this multiplication doesn't change the value of the number):

$$1,776 \times 10^0$$

Now, move the decimal point one place to the left and add 1 to the exponent until the decimal portion of the number is between 1 and 10:

$$= 177.6 \times 10^1$$
$$= 17.76 \times 10^2$$
$$= 1.776 \times 10^3$$

562. 9.008×10^5

Begin by multiplying 900,800 by 10^0:

$$900,800 \times 10^0$$

Now, move the decimal point one place to the left and add 1 to the exponent until the decimal portion of the number is between 1 and 10:

$$= 90,080 \times 10^1$$
$$= 9,008 \times 10^2$$
$$= 900.8 \times 10^3$$
$$= 90.08 \times 10^4$$
$$= 9.008 \times 10^5$$

563. 8.8199×10^2

Begin by multiplying 881.99 by 10^0:

$$881.99 \times 10^0$$

Now, move the decimal point one place to the left and add 1 to the exponent until the decimal portion of the number is between 1 and 10:

$$= 88.199 \times 10^1$$
$$= 8.8199 \times 10^2$$

564. 9.87654321×10^8

Begin by multiplying 987,654,321 by 10^0:

$$987,654,321 \times 10^0$$

Now, move the decimal point one place to the left and add 1 to the exponent until the decimal portion of the number is between 1 and 10 — that is, 8 places to the left:

$$= 9.87654321 \times 10^8$$

565. 1×10^7

Ten million is 10,000,000. Begin by multiplying 10,000,000 by 10^0:

$$10,000,000 \times 10^0$$

Now, move the decimal point one place to the left and add 1 to the exponent until the decimal portion of the number is between 1 and 10, but not 10 — that is, 7 places to the left:

$$= 1 \times 10^7$$

---

566. 4.1×10^{-1}

Begin by multiplying 0.41 by 10^0:

$$0.41 \times 10^0$$

Now, move the decimal point one place to the right and subtract 1 from the exponent until the decimal portion of the number is between 1 and 10:

$$= 4.1 \times 10^{-1}$$

---

567. 2.59×10^{-4}

Begin by multiplying 0.000259 by 10^0:

$$0.000259 \times 10^0$$

Now, move the decimal point one place to the right and subtract 1 from the exponent until the decimal portion of the number is between 1 and 10 — that is, 4 places to the right:

$$= 0.00259 \times 10^{-1}$$
$$= 0.0259 \times 10^{-2}$$
$$= 0.259 \times 10^{-3}$$
$$= 2.59 \times 10^{-4}$$

---

568. 1×10^{-3}

Begin by multiplying 0.001 by 10^0:

$$0.001 \times 10^0$$

Now, move the decimal point one place to the right and subtract 1 from the exponent until the decimal portion of the number is between 1 and 10 — that is, 3 places to the right:

$$= 1 \times 10^{-3}$$

---

569. 9×10^{-7}

Begin by multiplying 0.0000009 by 10^0:

$$0.0000009 \times 10^0$$

Now, move the decimal point one place to the right and subtract 1 from the exponent until the decimal portion of the number is between 1 and 10 — that is, 7 places to the right:

$$= 9 \times 10^{-7}$$

570. 1×10^{-6}

One-millionth written as a number is 0.000001. Begin by multiplying 0.000001 by 10^0:

$$0.000001 \times 10^0$$

Now, move the decimal point one place to the right and subtract 1 from the exponent until the decimal portion of the number is between 1 and 10 — that is, 6 places to the right:

$$= 1 \times 10^{-6}$$

571. **2,400**

Move the decimal point 3 places to the right and subtract 3 from the exponent:

$$= 2,400 \times 10^0$$

Now, drop the 10^0 entirely, because 10^0 equals 1:

$$= 2,400$$

572. **345,000**

Move the decimal point 5 places to the right and subtract 5 from the exponent:

$$= 345,000 \times 10^0$$

Now, drop the 10^0 entirely, because 10^0 equals 1:

$$= 345,000$$

573. **150,000,000 km**

Move the decimal point 8 places to the right and subtract 8 from the exponent:

$$= 150,000,000 \times 10^0$$

Now, drop the 10^0 entirely, because 10^0 equals 1:

$$= 150,000,000$$

574. **14.6 billion years**

Move the decimal point 10 places to the right and subtract 1 from the exponent:

$$= 14,600,000,000 \times 10^0$$

Now, drop the 10^0 entirely, because 10^0 equals 1:

$$= 14,600,000,000$$

This value is equal to 14.6 billion.

575. **31 trillion**

Move the decimal point 13 places to the right and subtract 13 from the exponent:

$$= 31{,}000{,}000{,}000{,}000 \times 10^0$$

Now, drop the 10^0 entirely, because 10^0 equals 1:

$$= 31{,}000{,}000{,}000{,}000$$

This value is equal to 31 trillion.

576. **0.075**

Move the decimal point 2 places to the left and add 2 to the exponent:

$$= 0.075 \times 10^0$$

Now, drop the 10^0 entirely, because 10^0 equals 1:

$$= 0.075$$

577. **3 thousandths**

Move the decimal point 3 places to the left and add 3 to the exponent:

$$= 0.003 \times 10^0$$

Now, drop the 10^0 entirely, because 10^0 equals 1:

$$= 0.003$$

This value is equivalent to 3 thousandths.

578. **0.0000254**

Move the decimal point 5 places to the left and add 5 to the exponent:

$$= 0.0000254 \times 10^0$$

Now, drop the 10^0 entirely, because 10^0 equals 1:

$$= 0.0000254$$

579. **0.0000000008**

Move the decimal point 10 places to the left and add 10 to the exponent:

$$= 0.0000000008 \times 10^0$$

Now, drop the 10^0 entirely, because 10^0 equals 1:

$$= 0.0000000008$$

Answers
501–600

580.

One ten-millionth

Move the decimal point 7 places to the left and add 7 to the exponent:

$$= 0.0000001 \times 10^0$$

Now, drop the 10^0 entirely, because 10^0 equals 1:

$$= 0.0000001$$

The digit 1 is in the ten millionths place.

581.

6×10^7

Multiply the decimal portions of the two values and multiply the powers of 10 by adding the exponents.

$$\left(2 \times 10^3\right) \times \left(3 \times 10^4\right)$$
$$= (2 \times 3) \times 10^{3+4}$$
$$= 6 \times 10^7$$

582.

7.7×10^8

Multiply the decimal portions of the two values and multiply the powers of 10 by adding the exponents.

$$\left(1.1 \times 10^6\right) \times \left(7 \times 10^2\right)$$
$$= (1.1 \times 7) \times 10^{6+2}$$
$$= 7.7 \times 10^8$$

583.

6.72×10^{10}

Multiply the decimal portions of the two values and multiply the powers of 10 by adding the exponents.

$$\left(1.6 \times 10^9\right) \times \left(4.2 \times 10^1\right)$$
$$= (1.6 \times 4.2) \times 10^{9+1}$$
$$= 6.72 \times 10^{10}$$

584.

8.785×10^3

Multiply the decimal portions of the two values and add the exponents:

$$\left(3.5 \times 10^{-4}\right) \times \left(2.51 \times 10^7\right)$$
$$= (3.5 \times 2.51) \times 10^{-4+7}$$
$$= 8.785 \times 10^3$$

585. 1.225×10^2

Multiply the decimal portions of the two values and multiply the powers of 10 by adding the exponents.

$$\left(2.5 \times 10^{-3}\right) \times \left(4.9 \times 10^4\right)$$
$$= \left(2.5 \times 4.9\right) \times 10^{-3+4}$$
$$= 12.25 \times 10^1$$

Now, move the decimal point one place to the left and add 1 to the exponent:

$$= 1.225 \times 10^2$$

586. 1.52×10^{-11}

Multiply the decimal portions of the two values and multiply the powers of 10 by adding the exponents.

$$\left(1.9 \times 10^{15}\right) \times \left(8 \times 10^{-27}\right)$$
$$= \left(1.9 \times 8\right) \times 10^{15+(-27)}$$
$$= 15.2 \times 10^{-12}$$

Now, move the decimal point one place to the left and add 1 to the exponent:

$$= 1.52 \times 10^{-11}$$

587. 3.417×10^1

Multiply the decimal portions of the two values and add the exponents:

$$\left(6.7 \times 10^1\right) \times \left(5.1 \times 10^{-1}\right)$$
$$= \left(6.7 \times 5.1\right) \times 10^{1+(-1)}$$
$$= 34.17 \times 10^0$$

Now, move the decimal point one place to the left and add 1 to the exponent:

$$= 3.417 \times 10^1$$

588. 2.5333×10^{21}

Multiply the decimal portions of the two values and add the exponents:

$$\left(3.29 \times 10^{20}\right) \times \left(7.7 \times 10^0\right)$$
$$= \left(3.29 \times 7.7\right) \times 10^{20+0}$$
$$= 25.333 \times 10^{20}$$

Now, move the decimal point one place to the left and add 1 to the exponent:

$$= 2.5333 \times 10^{21}$$

589. 7.4252533×10^7

Multiply the decimal portions of the three values and add the exponents:

$$\left(2.23 \times 10^7\right) \times \left(4.67 \times 10^{-9}\right) \times \left(7.13 \times 10^8\right)$$
$$= \left(2.23 \times 4.67 \times 7.13\right) \times 10^{7+(-9)+8}$$
$$= 74.252533 \times 10^6$$

Now, move the decimal point one place to the left and add 1 to the exponent:

$$= 7.4252533 \times 10^7$$

590. 3.4686×10^7

Multiply the decimal portions of the three values and add the exponents:

$$\left(9 \times 10^{-16}\right) \times \left(4.7 \times 10^{-24}\right) \times \left(8.2 \times 10^{45}\right)$$
$$= \left(9 \times 4.7 \times 8.2\right) \times 10^{-16+(-24)+45}$$
$$= 346.86 \times 10^5$$

Now, move the decimal point two places to the left and add 2 to the exponent:

$$= 3.4686 \times 10^7$$

591. 156

Convert 13 feet into inches by multiplying by 12:

$$13 \times 12 = 156$$

592. 1,080

Convert 18 hours into minutes by multiplying by 60:

$$18 \times 60 = 1,080$$

593. 240

Convert 15 pounds into ounces by multiplying by 16:

$$15 \times 16 = 240$$

594. 220

Convert 55 gallons into quarts by multiplying by 4:

$$55 \times 4 = 220$$

595. 190,080

First, convert 3 miles into feet by multiplying by 5,280:

$$3 \times 5,280 = 15,840$$

Next, convert 15,480 feet into inches by multiplying by 12:

$$15,840 \times 12 = 190,080$$

596. 416,000

First, convert 13 tons into pounds by multiplying by 2,000:

$$13 \times 2,000 = 26,000$$

Next, convert 26,000 pounds into ounces by multiplying by 16:

$$26,000 \times 16 = 416,000$$

597. 604,800

A week contains 7 days. To convert 7 days to hours, multiply 7 by 24:

$$24 \times 7 = 168$$

To convert 168 hours to minutes, multiply 168 by 60:

$$168 \times 60 = 10,080$$

To convert 10,080 minutes to seconds, multiply by 60:

$$10,080 \times 60 = 604,800$$

598. 2,176

First, convert 17 gallons into quarts by multiplying by 4:

$$17 \times 4 = 68$$

Next, convert 68 quarts into cups by multiplying by 4:

$$68 \times 4 = 272$$

Finally, convert 272 cups into fluid ounces by multiplying by 8:

$$272 \times 8 = 2,176$$

599. 46,112

First, convert 26.2 miles into feet by multiplying by 5,280:

$$26.2 \times 5,280 = 138,336$$

Next, convert 138,336 feet into yards by dividing by 3:

$$138,336 \div 3 = 46,112$$

600. 166,368,000

First, convert 5,199 tons into pounds by multiplying by 2,000:
$$5,199 \times 2,000 = 10,398,000$$

Next, convert 10,398,000 pounds into ounces by multiplying by 16:
$$10,398,000 \times 16 = 166,368,000$$

601. 2,522,880,000

A year contains 365 days. To convert 80 years to days, multiply 80 by 365:
$$80 \times 365 = 29,200$$

To convert 29,200 days to hours, multiply 29,200 by 24:
$$29,200 \times 24 = 700,800$$

To convert 700,800 hours to minutes, multiply by 60:
$$700,800 \times 60 = 42,048,000$$

To convert 42,048,000 minutes to seconds, multiply by 60:
$$42,048,000 \times 60 = 2,522,880,000$$

602. 11,520

A raindrop is $\frac{1}{90}$ fluid ounces, so a fluid ounce contains 90 raindrops. Multiply 90 by 8 to find the number of raindrops in a cup:
$$90 \times 8 = 720$$

Now, multiply 720 by 4 to find the number of raindrops in a quart:
$$720 \times 4 = 2,880$$

Finally, multiply 2,880 by 4 to find the number of raindrops in a gallon:
$$2,880 \times 4 = 11,520$$

603. 33

First, convert $53\frac{1}{3}$ yards into feet by multiplying by 3:
$$53\frac{1}{3} \times 3 = 160$$

Next, divide 5,280 by 160:
$$5,280 \div 160 = 33$$

604. 25,000

A liter contains 1,000 milliliters, so 25 liters contains 25,000 milliliters.

605. **800,000,000**

A megaton contains 1,000,000 tons, so 800 megatons contains 800,000,000 tons.

606. **30,000,000,000**

A second contains 1,000,000,000 (one billion) nanoseconds, so 30 seconds contains 30,000,000,000 (30 billion) nanoseconds.

607. **1,200,000**

A kilometer contains 1,000 meters, so 12 kilometers has 12,000 meters. And a meter contains 100 centimeters, so multiply 12,000 meters by 100:

$$12,000 \times 100 = 1,200,000$$

608. **17,000,000,000**

A megagram contains 1,000,000 grams, so 17 megagrams has 17,000,000 grams. And a gram contains 1,000 milligrams, so multiply 17,000,000 grams by 1,000:

$$17,000,000 \times 1,000 = 17,000,000,000$$

609. $\mathbf{9 \times 10^8}$

A gigawatt contains 1,000,000,000 watts, so 900 gigawatts has 900,000,000,000 watts. But it takes 1,000 watts to make up a kilowatt, so divide 900,000,000,000 watts by 1,000:

$$900,000,000,000 \div 1,000 = 900,000,000$$

To change this number to scientific notation, move the decimal point 8 places to the left and multiply by 10^8:

$$= 9 \times 10^8$$

610. $\mathbf{8.8 \times 10^{13}}$

A megadyne contains 1,000,000 dynes, so 88 megadynes has 88,000,000 dynes. And a dyne has 1,000,000 microdynes, so multiply 88,000,000 dynes by 1,000,000:

$$88,000,000 \times 1,000,000 = 88,000,000,000,000$$

To change this number to scientific notation, move the decimal point 10 places to the left and multiply by 10^{13}:

$$= 8.8 \times 10^{13}$$

611. $\mathbf{3.33 \times 10^{17}}$

A terameter contains 1 trillion meters (10^{12}), so multiply 333 by 10^{12}:

$$333 \times 10^{12}$$

A meter contains 1,000 millimeters (10^3), so multiply this result by 10^3:

$$333 \times 10^{12} \times 10^3 = 333 \times 10^{15}$$

Convert to scientific notation by moving the decimal point 2 places to the left and adding 2 to the exponent:

$$= 3.33 \times 10^{17}$$

612. **5.67811×10^{-1}**

A microsecond is one-millionth of a second, which is equivalent to 10^{-6}. Thus, multiply this amount by 567,811:

$$567,811 \times 10^{-6}$$

To convert to scientific notation, move the decimal point five places to the right and add 5 to the exponent:

$$= 5.67811 \times 10^{-1}$$

613. **10^{27}**

A nanogram contains 1,000 (10^3) picograms, and a gram contains 1 billion (10^9) nanograms, so multiply these two numbers to get the number of picograms in a gram:

$$10^3 \times 10^9 = 10^{12}$$

A teragram contains 1 trillion (10^{12}) grams, so multiply this number by the preceding result to get the number of picograms in a teragram:

$$10^{12} \times 10^{12} = 10^{24}$$

Finally, a petagram contains 1,000 (10^3) teragrams, so multiply this number by the preceding result to get the number of picograms in a petagram:

$$10^{24} \times 10^3 = 10^{27}$$

614. **5**

A kilobyte is 1,000 bytes, so a computer that can download 5 kilobytes of information in a nanosecond can download 5,000 bytes in a nanosecond. And a second contains 1 billion nanoseconds, so the number of bytes the computer can download in one second is

$$5,000 \times 1,000,000,000 = 5,000,000,000,000$$

However, there are 1 trillion bytes in a terabyte, so divide 5,000,000,000,000 by 1,000,000,000,000:

$$5,000,000,000,000 \div 1,000,000,000,000 = 5$$

615. **122°F**

Use the formula for converting Celsius to Fahrenheit:

$$F = (C \times 1.8) + 32 = (50 \times 1.8) + 32$$

Evaluate:

$$= 90 + 32 = 122$$

---

616. **37°C**

Use the formula for converting Fahrenheit to Celsius:

$$C = (F - 32) \div 1.8 = (98.6 - 32) \div 1.8$$

Evaluate:

$$= 66.6 \div 1.8 = 37$$

---

617. **22°C**

Use the formula for converting Fahrenheit to Celsius:

$$C = (F - 32) \div 1.8 = (72 - 32) \div 1.8$$

Evaluate:

$$= 40 \div 1.8 = 22.\overline{22} \approx 22$$

---

618. **58°C**

Use the formula for converting Fahrenheit to Celsius:

$$C = (F - 32) \div 1.8 = (136 - 32) \div 1.8$$

Evaluate:

$$= 104 \div 1.8 = 57.\overline{77} \approx 58$$

---

619. **2,795°F**

Use the formula for converting Celsius to Fahrenheit:

$$F = (C \times 1.8) + 32 = (1,535 \times 1.8) + 32$$

Evaluate:

$$2,763 + 32 = 2,795$$

---

620. **−459.67°F**

Use the formula for converting Celsius to Fahrenheit:

$$F = (C \times 1.8) + 32 = (-273.15 \times 1.8) + 32$$

Evaluate:

$$-491.67 + 32 = -459.67$$

621. **10 miles**

1 kilometer equals approximately one-half mile, so 1 mile equals approximately 2 kilometers. Multiply 20 by 2:

$$20 \div 2 = 10$$

622. **48 liters**

1 liter equals approximately $\frac{1}{4}$ gallon, so 1 gallon equals approximately 4 liters. Multiply 12 by 4:

$$12 \times 4 = 48$$

623. **90 kilograms**

1 kilogram equals approximately 2 pounds, so 1 pound equals approximately $\frac{1}{2}$ kilogram. Multiply 180 by $\frac{1}{2}$:

$$180 \times \frac{1}{2} = 90$$

624. **2,484 feet**

1 meter is approximately equal to 3 feet, so multiply 828 by 3:

$$828 \times 3 = 2,484$$

625. **20 meters**

1 meter equals approximately 3 feet, so 1 foot equals approximately $\frac{1}{3}$ meter. Multiply 60 by $\frac{1}{3}$:

$$60 \times \frac{1}{3} = 20$$

626. **10,000 pounds**

1 kilogram equals approximately 2 pounds, so multiply 5,000 by 2:

$$5,000 \times 2 = 10,000$$

627. **140 kilometers**

To begin, calculate the total distance in miles for 5 miles a day, 7 times per week, for 2 weeks:

$$5 \times 7 \times 2 = 70$$

Thus, the total distance is 70 miles. 1 kilometer is approximately equal to ½ mile, so multiply 70 by 2:

$$70 \times 2 = 140$$

---

628. **95 gallons**

First, calculate how many liters of gasoline the commuter puts in her car in 4 weeks by multiplying 95 by 4:

$$95 \times 4 = 380$$

One liter is approximately equal to $\frac{1}{4}$ gallon, so mulitiply 380 by $\frac{1}{4}$:

$$380 \times \frac{1}{4} = 95$$

---

629. **60 meters**

Begin by finding the length of the swimming pool in kilometers. To do this, multiply 2 (the number of kilometers in a mile) by $\frac{1}{32}$:

$$2 \times \frac{1}{32} = \frac{1}{16}$$

Thus, the swimming pool is $\frac{1}{16}$ kilometer in length. One kilometer is equal to 1,000 meters, so multiply $\frac{1}{16}$ by 1,000:

$$\frac{1}{16} \times 1,000 = 62.5$$

So rounded to the nearest 10 meters, the pool is approximately 60 meters.

---

630. **1.28 fluid ounces**

To begin, convert 40 milliliters to liters by dividing 40 by 1,000:

$$40 \div 1,000 = 0.04$$

One liter equals approximately 1 quart, so 0.04 liter equals approximately 0.04 quart. To convert 0.04 quart to cups, multiply by 4:

$$0.04 \times 4 = 0.16$$

To convert 0.16 cup to fluid ounces, multiply 0.16 by 8:

$$0.16 \times 8 = 1.28$$

---

631. **140**

The measures of two angles that result in a straight line always add up to 180 degrees. Thus, to find n, subtract as follows:

$$n = 180 - 40 = 140$$

---

632. **130**

When two lines intersect, the resulting vertical (opposite) angles are always equivalent. Therefore, $n = 130$.

633. 63

The measures of two angles that result in a straight line always add up to 180 degrees. Thus, to find n, subtract as follows:

$$n = 180 - 117 = 63$$

634. 61

The measures of three angles that result in a straight line always add up to 180 degrees. A right angle has 90 degrees, so to find n, subtract as follows:

$$n = 180 - 90 - 29 = 61$$

635. 14

The measures of three angles that result in a straight line always add up to 180 degrees. A square has four right angles, and a right angle measures 90 degrees. Thus, to find n, subtract as follows:

$$n = 180 - 90 - 76 = 14$$

636. 65

The measures of three angles of a triangle always add up to 180 degrees. Thus, to find n, subtract as follows:

$$n = 180 - 73 - 42 = 65$$

637. 91

The measures of two angles that result in a straight line always add up to 180 degrees. A square has four right angles, and a right angle measures 90 degrees. Thus, to find p, subtract as follows:

$$p = 180 - 158 = 22$$

The measures of the three angles of a triangle always add up to 180 degrees. Thus, to find n, subtract as follows:

$$n = 180 - 67 - 22 = 91$$

638. 14.5

The measures of the two smaller angles of a right triangle always add up to 90 degrees. Thus, to find n, subtract as follows:

$$n = 90 - 75.5 = 14.5$$

639. **61.6**

A rectangle has four right angles, each of which measures 90 degrees. Thus, to find n, subtract as follows:

$$n = 90 - 28.4 = 61.6$$

640. **75.4**

The measures of the four angles of a quadrilateral (four-sided polygon) always total 360 degrees. A right angle measures 90 degrees, so to find n, subtract as follows:

$$n = 360 - 90 - 108.2 - 86.4 = 75.4$$

641. **57.75**

When two lines are parallel, all corresponding angles are equivalent. Thus, you can determine the following:

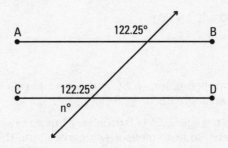

The measures of two angles that result in a straight line always add up to 180 degrees. Thus, to find n, subtract as follows:

$$n = 180 - 122.25 = 57.75$$

642. **86.6**

The measures of the five angles of a pentagon (five-sided polygon) always total 540 degrees. A right angle measures 90 degrees, so to find n, subtract as follows:

$$n = 540 - 90 - 118.3 - 83.9 - 161.2 = 86.6$$

643. **88.2**

The measures of two angles that result in a straight line always add up to 180 degrees. Thus, to find p, subtract as follows:

$$p = 180 - 134.1 = 45.9$$

An isosceles triangle has two equivalent angles, so you can draw the following:

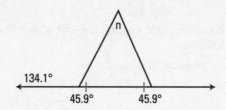

The measures of the three angles of a triangle always add up to 180 degrees. Thus, to find *n*, subtract as follows:

$$n = 180 - 45.9 - 45.9 = 88.2$$

644. **70.9**

When a triangle is inscribed in a circle such that one side of the triangle is a diameter of that circle, the opposite angle of that triangle is a right angle. Thus, *ABC* is a right triangle, so its two smaller angles add up to 90 degrees. Thus, to find *n*, subtract as follows:

$$n = 90 - 19.1 = 70.9$$

645. **69.75**

BCDE is a parallelogram, so $\overline{BC}$ and $\overline{ED}$ are parallel. Thus, angle *BCE* and angle *BEA* are equivalent, so angle *BEA* = 40.5.

$\overline{BE} = \overline{AE}$, so triangle *BEA* is isosceles. Thus the two remaining angles in this triangle are equivalent, so both measure *n* degrees. And the measures of the three angles in a triangle always add up to 180. Therefore, to find *n*, use the following equation:

$$180 = 40.5 + 2n$$
$$139.5 = 2n$$
$$69.75 = n$$

646. **36 square inches**

Use the formula for the area of a square:

$$A = s^2 = 6^2 = 36$$

647. **28 meters**

Use the formula for the perimeter of a square:

$$P = 4s = 4 \times 7 = 28$$

648. **10,201 square miles**

Use the formula for the area of a square:

$$A = s^2 = 101^2 = 10,201$$

649. **13.6 centimeters**

Use the formula for the perimeter of a square:

$$P = 4s = 4 \times 3.4 = 13.6$$

650. **21 feet**

Use the formula for the perimeter of a square, plugging in 84 for the perimeter; then solve for s.

$$P = 4s$$
$$84 = 4s$$
$$21 = s$$

651. **48 feet**

Begin by using the formula for the area of a square to find the side of the square. Plug in 144 for the area and solve for s.

$$A = s^2$$
$$144 = s^2$$
$$12 = s$$

Now, plug in 12 for s into the formula for the perimeter of a square:

$$P = 4s = 4 \times 12 = 48$$

652. **240.25 square feet**

Begin by using the formula for the perimeter of a square to find the side of the square. Plug in 62 for the perimeter and solve for s.

$$P = 4s$$
$$62 = 4s$$
$$15.5 = s$$

Now, plug 15.5 for s into the formula for the area of a square:

$$A = s^2 = 15.5^2 = 240.25$$

653. **60 feet**

Begin by plugging in 25 as the area into the formula for the area of a square ($A = s^2$) and solve for s.

$$25 = s^2$$
$$\sqrt{25} = \sqrt{s^2}$$
$$5 = s$$

Therefore, the side of the room is 5 yards. Convert yards to feet by multiplying by 3.

5 yards = 15 feet

Now, plug 15 into the formula for the perimeter of a square:

$$P = 4s = 4 \times 15 = 60$$

Therefore, the perimeter of the room is 60 feet.

654. 250,905,600 square feet

Begin by using the formula 1 mile = 5,280 feet to convert from miles to feet.

$$5,280 \times 3 = 15,840$$

Thus, the side of the square field is 15,840 feet. Plug this into the formula for the area of a square:

$$A = s^2 = 15,840^2 = 250,905,600$$

655. 0.4 kilometers

The perimeter of the park is 10 times greater than its area, so:

$$P = 10A$$

The perimeter of a square is 4s, so substitute this value for P into the equation above:

$$4s = 10A$$

The area of a square is s^2, so substitute this value for A into the preceding equation:

$$4s = 10s^2$$

To solve for s, begin by dividing both sides by s.

$$4 = 10s$$

Now, divide both sides by 10.

$$\frac{4}{10} = s, \text{ or } 0.4.$$

656. 24 square centimeters

Use the formula for the area of a rectangle:

$$A = lw = 8 \times 3 = 24$$

657. 36 meters

Use the formula for the perimeter of a rectangle.

$$P = 2l + 2w = (2 \times 16) + (2 \times 2)$$

Simplify.

$$= 32 + 4 = 36$$

---

658. **11.61 square feet**

Use the formula for the area of a rectangle.

$$A = lw = 4.3 \times 2.7 = 11.61$$

---

659. $3\frac{1}{4}$ **inches**

Use the formula for the perimeter of a rectangle.

$$P = 2l + 2w = \left(2 \times \frac{7}{8}\right) + \left(2 \times \frac{3}{4}\right)$$

Evaluate by canceling factors of 2:

$$= \frac{7}{4} + \frac{3}{2} = \frac{7}{4} + \frac{6}{4} = \frac{13}{4}$$

Convert this improper fraction to a mixed number:

$$= 3\frac{1}{4}$$

---

660. **155.25 square inches**

Use the formula for the area of a rectangle.

$$A = lw = 13.5 \times 11.5 = 155.25$$

---

661. $3\sqrt{10}$ **inches**

Use the formula for the area of a rectangle.

$$A = lw = \sqrt{15} \times \sqrt{6} = \sqrt{90}$$

Simplify by factoring.

$$= \sqrt{9}\sqrt{10} = 3\sqrt{10}$$

---

662. **50 feet**

Begin by using the formula for the area of a rectangle, plugging in 100 for the area and 5 for the width:

$$A = lw$$
$$100 = l \times 5$$

Divide both sides by 5.

$$20 = l$$

Thus, the length is 20. Now, use the formula for the perimeter of a rectangle, plugging in 20 for the length and 5 for the width.

$$P = 2l + 2w = (2 \times 20) + (2 \times 5)$$

Evaluate.

$$= 40 + 10 = 50$$

663. $23\frac{1}{2}$ inches

Begin by using the formula for the area of a rectangle, plugging in 30 for the area and 8 for the length.

$$A = lw$$
$$30 = 8 \times w$$

Divide both sides by 8.

$$\frac{30}{8} = w$$
$$\frac{15}{4} = w$$

Thus, the width is $\frac{15}{4}$. Now, use the formula for the perimeter of a rectangle, plugging in 8 for the length and $\frac{15}{4}$ for the width.

$$P = 2l + 2w = (2 \times 8) + \left(2 \times \frac{15}{4}\right)$$

Evaluate:

$$= 16 + \frac{15}{2} = 16 + 7\frac{1}{2} = 23\frac{1}{2}$$

664. 61 inches

Begin by using the formula for the area of a rectangle, plugging in 156 for the area and 24 for the length (because 2 feet = 24 inches).

$$A = lw$$
$$156 = 24 \times w$$

Divide both sides by 24.

$$6.5 = w$$

Thus, the width is 6.5. Now, use the formula for the perimeter of a rectangle, plugging in 24 for the length and 6.5 for the width.

$$P = 2l + 2w = (2 \times 24) + (2 \times 6.5)$$

Evaluate.

$$= 48 + 13 = 61$$

665. 24

If the area of a rectangle is 72 and both the length and width are whole numbers, you can write down all the possible lengths and widths as factor pairs of 72.

To begin, find all the factors of 72.

Factors of 72: 1, 2, 3, 4, 6, 8, 9, 12, 18, 24, 36, 72

$$72 \times 1$$
$$36 \times 2$$
$$24 \times 3$$
$$18 \times 4$$
$$12 \times 6$$
$$9 \times 8$$

Now, plug each of these pairs into the formula for the perimeter of a rectangle ($P = 2l + 2w$) until you find one that produces a perimeter of 54.

$$(2 \times 9) + (2 \times 8) = 18 + 16 = 34$$
$$(2 \times 12) + (2 \times 6) = 24 + 12 = 36$$
$$(2 \times 18) + (2 \times 4) = 36 + 8 = 44$$
$$(2 \times 12) + (2 \times 6) = 24 + 12 = 36$$
$$(2 \times 24) + (2 \times 3) = 48 + 6 = 54$$

Therefore, the length and width are 24 and 3.

666.   45

Use the formula for a parallelogram.

$$A = bh = 9 \times 5 = 45$$

667.   3,102.7

Use the formula for a parallelogram.

$$A = bh = 71 \times 43.7 = 3{,}102.7$$

668. $8\frac{13}{15}$

Use the formula for a parallelogram.

$$A = bh = 3\frac{4}{5} \times 2\frac{1}{3}$$

Evaluate by converting both mixed numbers to improper fractions and then multiplying.

$$= \frac{19}{5} \times \frac{7}{3} = \frac{133}{15} = 8\frac{13}{15}$$

669.   20

Use the formula for the area of a trapezoid.

$$A = \frac{b_1 + b_2}{2} h = \frac{7 + 3}{2} \times 4$$

Simplify the fraction.

$$= \frac{10}{2} \times 4 = 5 \times 4 = 20$$

670.　**85.32**

Use the formula for the area of a trapezoid.

$$A = \frac{b_1 + b_2}{2}h = \frac{9.2 + 12.4}{2} \times 7.9$$

Simplify the fraction.

$$= \frac{21.6}{2} \times 7.9 = 10.8 \times 7.9 = 85.32$$

671.　$\frac{23}{270}$

Use the formula for the area of a trapezoid.

$$A = \frac{b_1 + b_2}{2}h = \frac{\frac{1}{9} + \frac{2}{5}}{2} \times \frac{1}{3}$$

To simplify, begin by multiplying the two fractions.

$$= \frac{\frac{1}{9} + \frac{2}{5}}{6}$$

Next, add the two fractions in the numerator.

$$= \frac{\frac{5}{45} + \frac{18}{45}}{6} = \frac{\frac{23}{45}}{6}$$

Now, evaluate this fraction by turning it into fraction division.

$$= \frac{23}{45} \div 6 = \frac{23}{45} \times \frac{1}{6} = \frac{23}{270}$$

672.　**13.5 centimeters**

Use the formula for a parallelogram, plugging in 94.5 for the area and 7 for the base.

$$A = bh$$
$$94.5 = 7h$$

Divide both sides by 7.

$$13.5 = h$$

673.　**12**

Begin by plugging the area and bases into the formula for a trapezoid.

$$A = \frac{b_1 + b_2}{2}h$$
$$180 = \frac{9 + 21}{2}h$$

Simplify the fraction.

$$180 = \frac{30}{2}h$$
$$180 = 15h$$

Now, divide both sides by 15.

$$12 = h$$

---

674. $\frac{28}{45}$

Use the formula for a parallelogram, plugging in $\frac{4}{9}$ for the area and $\frac{5}{7}$ for the base.

$$A = bh$$
$$\frac{4}{9} = \frac{5}{7}h$$

Multiply both sides by $\frac{7}{5}$.

$$\frac{7}{5} \times \frac{4}{9} = \frac{7}{5} \times \frac{5}{7}h$$
$$\frac{28}{45} = h$$

---

675. 25.5

Begin by plugging the area, height, and base into the formula for a trapezoid.

$$A = \frac{b_1 + b_2}{2}h$$
$$45 = \frac{4.5 + b_2}{2}3$$

Divide both sides by 3, then multiply both sides by 2.

$$15 = \frac{4.5 + b_2}{2}$$
$$30 = 4.5 + b_2$$

Now, subtract 4.5 from both sides.

$$25.5 = b_2$$

---

676. 36 square inches

Use the formula for the area of a triangle ($A = \frac{1}{2}bh$) to solve the problem.

$$A = \frac{1}{2}bh = \frac{1}{2}(9)(8) = 36$$

---

677. 34.5 square meters

Use the formula for the area of a triangle ($A = \frac{1}{2}bh$) to solve the problem.

$$A = \frac{1}{2}bh = \frac{1}{2}(3)(23) = 34.5$$

678. $\dfrac{1}{12}$

Use the formula for the area of a triangle ($A = \frac{1}{2}bh$) to solve the problem.

$$A = \frac{1}{2}bh = \frac{1}{2}\left(\frac{4}{9}\right)\left(\frac{3}{8}\right)$$

Cancel common factors in the numerator and denominator.

$$= \frac{1}{1}\left(\frac{1}{3}\right)\left(\frac{1}{4}\right) = \frac{1}{12}$$

679. 110.5

Use the formula for the area of a triangle ($A = \frac{1}{2}bh$) to solve the problem.

$$A = \frac{1}{2}bh = \frac{1}{2}(17)(13) = 110.5$$

680. 99

In a right triangle, the lengths of the two legs (that is, the two short sides) are the base and height. Use the formula for the area of a triangle ($A = \frac{1}{2}bh$) to solve the problem.

$$A = \frac{1}{2}bh = \frac{1}{2}(9)(22) = 99$$

681. 24 square centimeters

In a right triangle, the lengths of the two legs (that is, the two short sides) are the base and height. Use the formula for the area of a triangle ($A = \frac{1}{2}bh$) to solve the problem.

$$A = \frac{1}{2}bh = \frac{1}{2}(4)(12) = 24$$

682. 30 meters

Use the formula for the area of a triangle ($A = \frac{1}{2}bh$) to solve the problem, plugging in 60 for the area and 4 for the height.

$$A = \frac{1}{2}bh$$

$$60 = \frac{1}{2}(b)(4)$$

To solve for the base b, first multiply $\frac{1}{2}$ by 4 on the right side of the equation; then solve for b.

$$60 = 2b$$

$$30 = b$$

683. 13

Use the formula for the area of a triangle ($A = \frac{1}{2}bh$) to solve the problem, plugging in 78 for the area. Be sure to convert the base to inches: 1 foot = 12 inches.

$$A = \frac{1}{2}bh$$

$$78 = \frac{1}{2}(12)(h)$$

To solve for the height h, first multiply $\frac{1}{2}$ by 12 on the right side of the equation, then solve for h by dividing both sides by 6.

$$78 = 6h$$

$$13 = h$$

684. $1\frac{3}{4}$ inches

Use the formula for the area of a triangle ($A = \frac{1}{2}bh$) to solve the problem, plugging in $\frac{5}{7}$ for the base and $\frac{5}{8}$ for the area.

$$A = \frac{1}{2}bh$$

$$\frac{5}{8} = \frac{1}{2}\left(\frac{5}{7}\right)(h)$$

To solve for the height h, first multiply $\frac{1}{2}$ by $\frac{5}{7}$ on the right side of the equation.

$$\frac{5}{8} = \frac{5}{14}h$$

Now, multiply both sides of the equation by $\frac{14}{5}$.

$$\frac{14}{5} \times \frac{5}{8} = \frac{14}{5} \times \frac{5}{14}h$$

$$\frac{14}{1} \times \frac{1}{8} = h$$

$$\frac{14}{8} = h$$

To finish, reduce the fraction $\frac{14}{8}$ and change it to a mixed number.

$$h = \frac{14}{8} = \frac{7}{4} = 1\frac{3}{4}$$

685. 13

To begin, use the formula for the area of a triangle ($A = \frac{1}{2}bh$), plugging in 84.5 for the area:

$$A = \frac{1}{2}bh$$

$$84.5 = \frac{1}{2}bh$$

Multiply both sides by 2 to get rid of the fraction.

$$169 = bh$$

The base and height are the same, so you can use the same variable h for both of these values. Therefore, $bh = (h)h = h^2$, so you can substitute h^2 for bh in the preceding equation:

$$169 = h^2$$

Solve for h by taking the square root of both sides.

$$\sqrt{169} = \sqrt{h^2}$$
$$13 = h$$

686.   **5 feet**

Use the Pythagorean Theorem ($a^2 + b^2 = c^2$) to find the hypotenuse:

$$a^2 + b^2 = c^2$$
$$3^2 + 4^2 = c^2$$
$$9 + 16 = c^2$$
$$25 = c^2$$
$$\sqrt{25} = \sqrt{c^2}$$
$$5 = c$$

687.   **26 centimeters**

Use the Pythagorean Theorem ($a^2 + b^2 = c^2$) to find the hypotenuse:

$$a^2 + b^2 = c^2$$
$$10^2 + 24^2 = c^2$$
$$100 + 576 = c^2$$
$$676 = c^2$$

To finish, take the square root of both sides of the equation:

$$\sqrt{676} = \sqrt{c^2}$$
$$26 = c$$

688. $4\sqrt{5}$

Use the Pythagorean Theorem ($a^2 + b^2 = c^2$) to find the hypotenuse:

$$a^2 + b^2 = c^2$$
$$4^2 + 8^2 = c^2$$
$$16 + 64 = c^2$$
$$80 = c^2$$

Take the square root of both sides of the equation.

$$\sqrt{80} = \sqrt{c^2}$$
$$\sqrt{80} = c$$

Simplify by factoring as follows:

$$\sqrt{16}\sqrt{5} = c$$
$$4\sqrt{5} = c$$

---

689. $\sqrt{265}$

Use the Pythagorean Theorem $(a^2 + b^2 = c^2)$ to find the hypotenuse:

$$a^2 + b^2 = c^2$$
$$11^2 + 12^2 = c^2$$
$$121 + 144 = c^2$$
$$265 = c^2$$

Take the square root of both sides of the equation.

$$\sqrt{265} = \sqrt{c^2}$$
$$\sqrt{265} = c$$

---

690. $\sqrt{5}$

Use the Pythagorean Theorem $(a^2 + b^2 = c^2)$ to find the hypotenuse:

$$a^2 + b^2 = c^2$$
$$\sqrt{2}^2 + \sqrt{3}^2 = c^2$$
$$2 + 3 = c^2$$
$$5 = c^2$$

Take the square root of both sides of the equation.

$$\sqrt{5} = \sqrt{c^2}$$
$$\sqrt{5} = c$$

---

691. $\sqrt{173}$

Use the Pythagorean Theorem $(a^2 + b^2 = c^2)$ to find the hypotenuse:

$$a^2 + b^2 = c^2$$
$$\left(4\sqrt{3}\right)^2 + \left(5\sqrt{5}\right)^2 = c^2$$

Evaluate the left side of the equation.

$$\left(4^2\sqrt{3}^2\right) + \left(5^2\sqrt{5}^2\right) = c^2$$
$$(16 \times 3) + (25 \times 5) = c^2$$
$$48 + 125 = c^2$$
$$173 = c^2$$

Take the square root of both sides of the equation.

$$\sqrt{173} = \sqrt{c^2}$$

$$\sqrt{173} = c$$

692.   1

Use the Pythagorean Theorem $(a^2 + b^2 = c^2)$ to find the hypotenuse:

$$a^2 + b^2 = c^2$$

$$\left(\frac{5}{13}\right)^2 + \left(\frac{12}{13}\right)^2 = c^2$$

Evaluate the left side of the equation using the following steps:

$$\left(\frac{5}{13}\right)\left(\frac{5}{13}\right) + \left(\frac{12}{13}\right)\left(\frac{12}{13}\right) = c^2$$

$$\frac{25}{169} + \frac{144}{169} = c^2$$

$$\frac{169}{169} = c^2$$

$$1 = c^2$$

$$\sqrt{1} = \sqrt{c^2}$$

$$1 = c$$

693. $\frac{5}{12}$

Use the Pythagorean Theorem $(a^2 + b^2 = c^2)$ to find the hypotenuse:

$$a^2 + b^2 = c^2$$

$$\left(\frac{1}{3}\right)^2 + \left(\frac{1}{4}\right)^2 = c^2$$

Evaluate the left side of the equation.

$$\left(\frac{1}{3}\right)\left(\frac{1}{3}\right) + \left(\frac{1}{4}\right)\left(\frac{1}{4}\right) = c^2$$

$$\frac{1}{9} + \frac{1}{16} = c^2$$

$$\frac{25}{144} = c^2$$

Take the square root of both sides of the equation.

$$\sqrt{\frac{25}{144}} = \sqrt{c^2}$$

$$\frac{\sqrt{25}}{\sqrt{144}} = \sqrt{c^2}$$

$$\frac{5}{12} = c$$

694. **40**

Use the Pythagorean Theorem ($a^2 + b^2 = c^2$) to find the length of the longer leg:
$$a^2 + b^2 = c^2$$
$$75^2 + b^2 = 85^2$$
$$5,625 + b^2 = 7,225$$

Subtract 5,625 from both sides, then take the square root of both sides of the equation.
$$b^2 = 1,600$$
$$\sqrt{b^2} = \sqrt{1,600}$$
$$b = 40$$

695. **$7\sqrt{3}$**

Use the Pythagorean Theorem ($a^2 + b^2 = c^2$) to find the length of the longer leg:
$$a^2 + b^2 = c^2$$
$$7^2 + b^2 = 14^2$$
$$49 + b^2 = 196$$

Subtract 49 from both sides, then take the square root of both sides of the equation.
$$b^2 = 147$$
$$\sqrt{b^2} = \sqrt{147}$$
$$b = \sqrt{147}$$

Simplify by factoring as follows:
$$b = \sqrt{49}\sqrt{3}$$
$$b = 7\sqrt{3}$$

696. **16**

Use the formula for the diameter of a circle:
$$D = 2r = 2 \times 8 = 16$$

697. **121π**

Use the formula for the area of a circle:
$$A = \pi r^2 = \pi \times 11^2 = 121\pi$$

698. **40π**

Use the formula for the circumference of a circle:
$$C = 2\pi r = 2 \times \pi \times 20 = 40\pi$$

699. 2.89π

Use the formula for the area of a circle:

$$A = \pi r^2 = \pi \times 1.7^2 = 2.89\pi$$

700. 13π

Use the formula for the circumference of a circle:

$$C = 2\pi r = 2 \times \pi \times 6.5 = 13\pi$$

701. 99π

The formula for the diameter of a circle is $D = 2r$, and the formula for the circumference is $C = 2\pi r$. Notice that the only difference between the diameter of a circle ($D = 2r$) and its circumference ($C = 2\pi r$) is a factor of π. So, a quick way to change the diameter to the circumference is simply to multiply by π.

Therefore, if a circle has a diameter of 99, its circumference is 99π.

702. $\dfrac{5}{6}\pi$

The formula for the diameter of a circle is $D = 2r$, and the formula for the circumference is $C = 2\pi r$. Notice that the only difference between the diameter of a circle ($D = 2r$) and its circumference ($C = 2\pi r$) is a factor of π. So, a quick way to change the diameter to the circumference is simply to multiply by π.

Therefore, if a circle has a diameter of $\dfrac{5}{6}$, its circumference is $\dfrac{5}{6}\pi$.

703. $2,500\pi$

A circle with a diameter of 100 has a radius of 50 (because $D = 2r$). Plug this value into the formula for the area of a circle:

$$A = \pi r^2 = \pi \times 50^2 = 2,500\pi$$

704.   9

Use the formula for the area of a circle, plugging in 81π for the area:

$$A = \pi r^2$$
$$81\pi = \pi r^2$$

Divide both sides of the equation by π.

$$81 = r^2$$

Now, take the square root of each side.

$$\sqrt{81} = \sqrt{r^2}$$
$$9 = r$$

705. **33**

Use the formula for the circumference of a circle, plugging in 66π for the circumference:

$$C = 2\pi r$$
$$66\pi = 2\pi r$$

Divide both sides of the equation by π and then by 2.

$$66 = 2r$$
$$33 = r$$

706. **29.16π**

To begin, find the radius using the formula for the circumference of a circle, plugging in 10.8π for the circumference:

$$C = 2\pi r$$
$$10.8\pi = 2\pi r$$

Divide both sides of the equation by π and then by 2.

$$10.8 = 2r$$
$$5.4 = r$$

Now, use the area formula, plugging in 5.4 for the radius:

$$A = \pi r^2 = \pi \times 5.4^2 = 29.16\pi$$

707. $\frac{4}{5}\pi$

To begin, find the radius using the formula for the area of a circle, plugging in $\frac{4}{25}\pi$ for the area:

$$A = \pi r^2$$
$$\frac{4}{25}\pi = \pi r^2$$

Divide both sides of the equation by π.

$$\frac{4}{25} = r^2$$

Now, take the square root of both sides.

$$\sqrt{\frac{4}{25}} = \sqrt{r^2}$$
$$\frac{\sqrt{4}}{\sqrt{25}} = \sqrt{r^2}$$
$$\frac{2}{5} = r$$

Now, use the circumference formula, plugging in $\frac{2}{5}$ for the radius:

$$C = 2\pi r = 2 \times \pi \times \frac{2}{5} = \frac{4}{5}\pi$$

708. $\dfrac{4}{\sqrt{\pi}}$

Use the formula for the area of a circle, plugging in 16 for the area:

$$A = \pi r^2$$
$$16 = \pi r^2$$

Divide both sides of the equation by π.

$$\frac{16}{\pi} = r^2$$

Now, take the square root of both sides.

$$\sqrt{\frac{16}{\pi}} = \sqrt{r^2}$$
$$\frac{\sqrt{16}}{\sqrt{\pi}} = \sqrt{r^2}$$
$$\frac{4}{\sqrt{\pi}} = r$$

709. $\dfrac{85.5625}{\pi}$

To begin, find the radius using the formula for the circumference of a circle, plugging in 18.5 for the circumference:

$$C = 2\pi r$$
$$18.5 = 2\pi r$$

Divide both sides of the equation by 2 and then by π.

$$9.25 = \pi r$$
$$\frac{9.25}{\pi} = r$$

Now, use the area formula, plugging in $\dfrac{9.25}{\pi}$ for the radius:

$$A = \pi r^2 = \pi \times \left(\frac{9.25}{\pi}\right)^2$$

To finish, first evaluate the power.

$$= \pi \times \frac{9.25^2}{\pi^2} = \frac{85.5625\pi}{\pi^2}$$

Now, cancel a factor of π in both the numerator and denominator.

$$= \frac{85.5625}{\pi}$$

710. **1,728 cubic inches**

Use the formula for the volume of a cube:

$$V = s^3 = 12^3$$

Evaluate as follows:

$$= 12 \times 12 \times 12 = 1,728$$

711. **421.875**

Use the formula for the volume of a cube:

$$V = s^3 = 7.5^3$$

Evaluate as follows:

$$= 7.5 \times 7.5 \times 7.5 = 421.875$$

712. **100 inches**

Use the formula for the volume of a cube, plugging in 1,000,000 for the volume:

$$V = s^3$$
$$1,000,000 = s^3$$

To find the value of s, you want to find a number which, when multiplied by itself 3 times, equals 1,000,000. A little trial and error makes this obvious:

$$10 \times 10 \times 10 = \cancel{1,000}$$
$$100 \times 100 \times 100 = 1,000,000$$

713. **600 cubic inches**

Use the formula for the volume of a box:

$$V = lwh = 15 \times 4 \times 10 = 600$$

714. **327.25 cubic inches**

Use the formula for the volume of a box:

$$V = lwh = 8.5 \times 11 \times 3.5 = 327.25$$

715. $\dfrac{55}{512}$ **cubic inches**

Use the formula for the volume of a box:

$$V = lwh = \frac{1}{4} \times \frac{5}{8} \times \frac{11}{16} = \frac{55}{512}$$

716. **5 centimeters**

Use the formula for a box, plugging in the value 20,000 for the volume, 80 for the length, and 50 for the width:

$$V = lwh$$
$$20,000 = 80 \times 50 \times h$$

Simplify and divide both sides by 4,000.

$$20,000 = 4,000h$$
$$5 = h$$

717. **0.0456 inches**

Use the formula for a box, plugging in the value 45.6 for the volume, 10 for the length, and 100 for the height:

$$V = lwh$$
$$45.6 = 10 \times w \times 100$$

Simplify and divide both sides by 1,000.

$$45.6 = 1,000 \times w$$
$$0.0456 = w$$

718. **24π cubic feet**

Use the formula for the volume of a cylinder:

$$V = \pi r^2 h = \pi \times 2^2 \times 6$$

Simplify.

$$= \pi \times 4 \times 6 = 24\pi$$

719. **$222,750\pi$**

Use the formula for the volume of a cylinder:

$$V = \pi r^2 h = \pi \times 45^2 \times 110$$

Evaluate.

$$= \pi \times 2,025 \times 110 = 222,750\pi$$

720. **0.176π cubic meters**

Use the formula for the volume of a cylinder:

$$V = \pi r^2 h = \pi \times 0.4^2 \times 1.1$$

Simplify:

$$= \pi \times 0.16 \times 1.1 = 0.176\pi$$

721. **$\frac{45}{128}\pi$ cubic inches**

Use the formula for the volume of a cylinder:

$$V = \pi r^2 h = \pi \times \left(\frac{3}{4}\right)^2 \times \frac{5}{8}$$

Simplify.

$$= \pi \times \left(\frac{3}{4}\right)\left(\frac{3}{4}\right) \times \frac{5}{8} = \pi \times \frac{9}{16} \times \frac{5}{8} = \frac{45}{128}\pi$$

722.   **6.5 feet**

Use the formula for the volume of a cylinder, plugging in 58.5π for the volume and 3 for the radius:

$$V = \pi r^2 h$$
$$58.5\pi = \pi \times 3^2 \times h$$
$$58.5\pi = \pi \times 9 \times h$$

Divide both sides by π and then by 9.

$$58.5 = 9h$$
$$6.5 = h$$

723.   **36π**

Use the formula for the volume of a sphere:

$$V = \frac{4}{3}\pi r^3 = \frac{4}{3} \times \pi \times 3^3 = \frac{4 \times \pi \times 27}{3}$$

Cancel a factor of 3 in both the numerator and denominator and then simplify.

$$= 4 \times \pi \times 9 = 36\pi$$

724.   **$\frac{9}{128}\pi$**

Use the formula for the volume of a sphere:

$$V = \frac{4}{3}\pi r^3 = \frac{4}{3} \times \pi \times \left(\frac{3}{8}\right)^3$$

Evaluate the power, cancel factors where possible, and then multiply:

$$= \frac{4}{3} \times \pi \times \left(\frac{3}{8}\right)\left(\frac{3}{8}\right)\left(\frac{3}{8}\right) = \frac{1}{1} \times \pi \times \left(\frac{1}{2}\right)\left(\frac{3}{8}\right)\left(\frac{3}{8}\right) = \frac{9}{128}\pi$$

725.   **2.304π cubic meters**

Use the formula for the volume of a sphere:

$$V = \frac{4}{3}\pi r^3 = \frac{4}{3} \times \pi \times 1.2^3$$

Evaluate the power.

$$= \frac{4}{3} \times \pi \times (1.2)(1.2)(1.2) = \frac{4}{3} \times \pi \times 1.728$$

Now, multiply $\frac{4}{3}$ by 1.728. You can do this in two steps: First multiply by 4 and then divide by 3:

$$= \frac{6.912}{3} \times \pi$$
$$= 2.304\pi$$

726. $\frac{1}{2}$ **foot**

Using the formula for a sphere, plug in $\frac{1}{6}\pi$ as the volume:

$$V = \frac{4}{3}\pi r^3$$

$$\frac{1}{6}\pi = \frac{4}{3}\pi r^3$$

Divide both sides of the equation by π.

$$\frac{1}{6} = \frac{4}{3}r^3$$

Now, multiply both sides of the equation by $\frac{3}{4}$.

$$\frac{3}{4} \times \frac{1}{6} = \frac{3}{4} \times \frac{4}{3}r^3$$

$$\frac{1}{8} = r^3$$

The radius r is a number which, when multiplied by itself 3 times, equals $\frac{1}{8}$. This number is $\frac{1}{2}$, because:

$$\frac{1}{2} \times \frac{1}{2} \times \frac{1}{2} = \frac{1}{8}$$

727. **32 cubic inches**

Use the formula for the volume of a pyramid:

$$V = \frac{1}{3}s^2h = \frac{1}{3} \times 4^2 \times 6$$

Evaluate the power and cancel a factor of 3 in the numerator and denominator.

$$= \frac{1}{3} \times 16 \times 6 = 16 \times 2 = 32$$

728. **4 meters**

Use the formula for the volume of a pyramid, plugging in 80 for the volume and 15 for the height:

$$V = \frac{1}{3}s^2h$$

$$80 = \frac{1}{3} \times s^2 \times 15$$

Cancel a factor of 3 in the numerator and denominator; then divide both sides of the equation by 5.

$$80 = s^2 \times 5$$

$$16 = s^2$$

Now, take the square root of both sides of the equation.

$$\sqrt{16} = \sqrt{s^2}$$

$$4 = s$$

729. **$3{,}000\pi$ cubic inches**

Use the formula for the volume of a cone:

$$V = \frac{1}{3}\pi r^2 h = \frac{1}{3}\pi \times 30^2 \times 10$$

Evaluate the power and cancel a factor of 3 in the numerator and denominator.

$$= \frac{1}{3}\pi \times 900 \times 10 = 300\pi \times 10 = 3{,}000\pi$$

730. **11**

Use the formula for the volume of a cone, plugging in 132π for the volume and 6 for the radius:

$$V = \frac{1}{3}\pi r^2 h$$

$$132\pi = \frac{1}{3} \times \pi \times 6^2 \times h$$

Evaluate the power, cancel a factor of 3 in the numerator and denominator, and then divide both sides of the equation by 12π.

$$132\pi = \frac{1}{3} \times \pi \times 36 \times h$$

$$132\pi = 12\pi h$$

$$11 = h$$

731. **Kent**

Brian collected $300 and Kent collected $500, so Kent collected $200 more than Brian.

732. **$1,800**

Arianna collected $600, Eva collected $800, and Stella collected $400. Therefore, together they collected $600 + $800 + $400 = $1,800.

733. $\frac{1}{9}$

Stella collected $400. The total amount collected was $600 + $300 + $800 + $500 + $400 + $1,000 = $3,600. Make a fraction of these two amounts as follows:

$$\frac{Stella}{Total} = \frac{\$400}{\$3{,}600} = \frac{1}{9}$$

734. **2:5**

Stella collected $400 and Tyrone collected $1,000. To find the ratio, make a fraction of these two numbers and reduce it:

$$\frac{\$400}{\$1{,}000} = \frac{4}{10} = \frac{2}{5}$$

Therefore, Stella and Tyrone collected funds in a 2:5 ratio.

735. **Kent**

Eva collected $800, so if she had collected $300 less, she would have collected $500. Kent collected $500.

736. **44%**

Arianna collected $600 and Tyrone collected $1,000, so together they collected $1,600. The total amount collected was $3,600 (see Answer 733). Make a fraction of these two amounts:

$$\frac{\$1,600}{\$3,600} = \frac{4}{9}$$

Now, divide $4 \div 9$ to convert this fraction into a repeating decimal and then into a percent:

$$= 0.\overline{4} = 44.\overline{4}\% \approx 44\%$$

737. **Biochemistry and Economics**

Biochemistry accounts for 35% of Kaitlin's study time, and Economics accounts for 15%. Together, these account for 50% of her time.

738. **Calculus, Economics, and Spanish**

Calculus accounts for 20% of Kaitlin's study time, Economics 15%, and Spanish 20%. Together, these account for 55% of her time.

739. **4 hours**

Kaitlin spends 20% of her time studying for Spanish. Thus, if she spent 20 hours last week studying, she spent 20% of 20 hours studying for Spanish:

$$0.2 \times 20 = 4$$

Therefore, she spent 4 hours studying for Spanish.

740. **30 hours**

Kaitlin spent 20% of her time studying for Calculus and 15% of her time studying for Economics. Thus, she spent 5% more time studying for Calculus than Economics. So if 5% of her time represented 1.5 hours, multiplying this value by 20 would represent 100% of her studying time (because 5% times 20 = 100%):

$$1.5 \times 20 = 30$$

Therefore, Kaitlin spent 30 hours studying.

741. **30 hours**

Kaitlin spends 10% of her time studying for Physics. Thus, if she spent 3 hours studying for this class, she spent 10 times more than that studying for all of her classes. Therefore, she spent 30 hours studying for all of her classes.

742. **4 hours and 40 minutes**

Kaitlin spends 15% of her time studying for Economics. If this accounted for 2 hours, then 1/3 of this time – that is, 40 minutes – would account for 5% of her time. Then, multiplying this amount of time by 7 (40 minutes × 7 = 280 minutes) would account for 35% of her time. Thus, Kaitlin spent 280 minutes studying for biochemistry, which equals 4 hours and 40 minutes.

743. **October**

Net profit was $2,800 in February and the same in October.

744. **$8,800**

Net profit for January, February, and March was $2,400 + $2,800 + $3,600 = $8,800.

745. **March**

Between March and April, net profit increased by $4,400 – $3,600 = $800. This is equivalent to the profit shown in March when compared with February ($800), but greater than the increase in profit shown in February ($400), May (decrease in profit), June ($400), July ($400), August ($400), September and October (decrease in profit), November ($400) or December ($400).

746. **August and September**

In August and September, the combined net profit was $5,200 + $3,600 = $8,800.

747. **January**

To begin, calculate the total net profit for the year:

$2,400 + $2,800 + $3,600 + $4,400 + $4,000 + 4,400 + $4,800 + $5,200 + $ 3,600 + $2,800 + $3,200 + $3,600 = $44,800

Now, calculate 5% of $44,800:

$44,800 × 0.05 = $2,240

The nearest net profit to $2,240 was $2,400, in January.

748. 22,000

Plattfield is the largest town in Alabaster County. Its population is equivalent to 11 stick figures, each of which represents 2,000 people, so its population is $2,000 \times 11 = 22,000$.

749. Talkingham

To begin, find the total population of the county:

$$9,000 + 12,000 + 15,000 + 22,000 + 6,000 + 14,000 = 78,000$$

Now, calculate $\frac{1}{6}$ of 78,000:

$$\frac{1}{6} \times 78,000 = 78,000 \div 6 = 13,000$$

Talkingham has a population of 14,000, which is slightly more than 13,000.

750. 19%

As calculated in Answer 749, the entire county has a population of 78,000. Morrissey Station has a population of 15,000. Make a fraction of these two numbers as follows:

$$\frac{15,000}{78,000} = \frac{5}{26}$$

Change this fraction to a decimal by dividing $5 \div 26$, and then change the decimal to a percent as follows:

$$5 \div 26 \approx 0.192 = 19\%$$

751. Barker Lake and Talkingham

Plattfield has a population of 22,000 people. Barker Lake has a population of 9,000 and Talkingham has a population of 14,000. Therefore, together, Barker Lake and Talkingham have a combined population of $9,000 + 14,000 = 23,000$, which is 1,000 greater than the population of Plattfield.

752. 20%

The population of Talkingham is 14,000. If it increased by 2,000 (one stick figure), then its population would be 16,000. And if all the other towns remained constant in their population, then the population of the county would also rise by 2,000 people, from 78,000 (see Answer 749) to 80,000.

Make a fraction from these two numbers:

$$\frac{16,000}{80,000} = \frac{1}{5} = 20\%$$

753. 69%

The two top candidates were Bratlafski with 41% and McCullers with 28%, so together they received 69%.

754.

Farelese and McCullers

Faralese received 7% of the vote and McCullers received 28%, so together they received 7% + 28% = 35%.

755.

3,000

Faralese received 7% of the vote, and Williamson received 4%. If 100,000 votes were cast, Faralese received 7,000 votes (0.07 × 100,000) and Williamson received 4,000 (0.04 × 100,000). Therefore, Faralese received 3,000 more votes than Williamson.

756.

200,000

Jordan received 17% of the vote, so if she had received 34,000 votes, each percentage point would count for:

$$\frac{34,000}{17} = 2,000$$

Thus, if each percentage point counted for 2,000 votes, 100% of the vote would be 200,000 votes.

757.

39,200

Bratlaski received 41% of the vote and Pardee received 3%. Thus, Bratlaski received 38% more votes than Pardee. If 38% of the vote represented 53,200 votes, each percentage point would count for:

$$\frac{53,200}{38} = 1,400$$

Thus, if each percentage point counted for 1,400 votes, McCullers' share of 28% of the vote would be:

$$1,400 \times 28 = 39,200$$

758.

2,500

Seven hundred and fifty trees were planted in Edinburgh County and 1,750 in Manchester County, so together there were 750 + 1,750 = 2,500 trees.

759.

8.000

The total number of trees was as follows:

$$1,500 + 500 + 2,250 + 750 + 1,250 + 1,750 = 8,000$$

760.

Dublin and Manchester

In Answer 759, the total number of trees among the six counties is calculated at 8,000. Thus, 50% of the trees is 4,000. Dublin accounts for 2,250 trees and Manchester accounts for 1,750, so together this accounts for 2,500 + 1,750 = 4,000.

761. **Birmingham**

The total number of trees was 8,000 (see Answer 759). Thus, 18.75% of the trees is

$$0.1875 \times 8,000 = 1,500$$

Fifteen hundred trees were planted in Birmingham County.

762. $\frac{1}{6}$

The total number of trees was 8,000 (see Answer 759), of which 500 were in Calais County. If 1,000 additional trees had been planted in Calais County, then 1,500 trees would have been planted there out of a total of 9,000. Make a fraction from these two numbers and reduce:

$$\frac{1,500}{9,000} = \frac{15}{90} = \frac{1}{6}$$

763. **See below.**

i. Q

ii. S

iii. R

iv. P

v. T

764. **6**

$Q = (1, 6)$. To go from $(0, 0)$ to $(1, 6)$, you need to go

up 6, over 1

Translate these words as follows:

+ 6 / 1

Thus, the slope of the line that passes through both the origin and Q is

$$+\frac{6}{1} = 6$$

765. $\frac{1}{3}$

$S = (-3, -1)$. To go from $(-3, -1)$ to $(0, 0)$, you need to go

up 1, over 3

Translate these words as follows:

+ 1 / 3

Thus, the slope of the line that passes through both the origin and S is

$$+\frac{1}{3} = \frac{1}{3}$$

766. −1

$P = (3, 4)$ and $Q = (1, 6)$. To go from $(1, 6)$ to $(3, 4)$, you need to go

down 2, over 2

Translate these words as follows:

$-2 / 2$

Thus, the slope of the line that passes through both P and Q is

$$-\frac{2}{2} = -1$$

767. $-\frac{8}{7}$

$R = (-2, 5)$ and $T = (5, -3)$. To go from $(-2, 5)$ to $(5, -3)$, you need to go

down 8, over 7

Translate these words as follows:

$-8 / 7$

Thus, the slope of the line that passes through both R and T is

$$-\frac{8}{7}$$

768. $-\frac{1}{4}$

$S = (-3, -1)$, $T = (5, -3)$. To go from $(-3, -1)$ to $(5, -3)$, you need to go

down 2, over 8

Translate these words as follows:

$-2 / 8$

Thus, the slope of the line that passes through both R and T is

$$-\frac{2}{8} = -\frac{1}{4}$$

769. 5

To begin, draw a right triangle with the line that you want to measure as the hypotenuse:

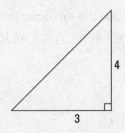

Now, notice that the horizontal leg of this triangle has a length of 3, and the vertical leg has a length of 4. Thus, this is a 3-4-5 right triangle, so the distance between the origin and *P* is 5.

770. $\sqrt{37}$

To begin, draw a right triangle with the line that you want to measure as the hypotenuse:

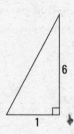

Now, notice that the horizontal leg of this triangle has a length of 1, and the vertical leg has a length of 6. Use the Pythagorean theorem to measure the length of the hypotenuse:

$$a^2 + b^2 = c^2$$
$$1^2 + 6^2 = c^2$$
$$1 + 36 = c^2$$
$$37 = c^2$$
$$\sqrt{37} = c$$

Thus, the distance between *R* and *S* is $\sqrt{37}$.

771. 8

To find the average, use the following formula:

$$Mean = \frac{Sum\ of\ items}{Number\ of\ items} = \frac{4+9+11}{3}$$

Simplify.

$$= \frac{24}{3} = 8$$

772. 26

To find the average, use the following formula:

$$Mean = \frac{Sum\ of\ items}{Number\ of\ items} = \frac{2+2+16+29+81}{5}$$

Simplify.

$$= \frac{130}{5} = 26$$

773.   **1,411**

To find the average, use the following formula:

$$Mean = \frac{Sum\ of\ items}{Number\ of\ items} = \frac{245 + 1,024 + 2,964}{3}$$

Simplify.

$$= \frac{4,233}{3} = 1,411$$

774.   **48.2**

To find the average, use the following formula:

$$Mean = \frac{Sum\ of\ items}{Number\ of\ items} = \frac{17 + 23 + 35 + 64 + 102}{5}$$

Simplify.

$$= \frac{241}{5} = 48.2$$

775.   **9.1**

To find the average, use the following formula:

$$Mean = \frac{Sum\ of\ items}{Number\ of\ items} = \frac{3.5 + 4.1 + 9.2 + 19.6}{4}$$

Simplify.

$$= \frac{36.4}{4} = 9.1$$

776.   **307.418**

To find the average, use the following formula:

$$Mean = \frac{Sum\ of\ items}{Number\ of\ items} = \frac{7.214 + 91.8 + 823.24}{3}$$

Simplify.

$$= \frac{922.254}{3} = 307.418$$

777. $\frac{7}{45}$

To begin, find the sum of $\frac{1}{5}$ and $\frac{1}{9}$.

$$\frac{1}{5} + \frac{1}{9} = \frac{14}{45}$$

Now, plug this result into the numerator of the formula for the mean, with 2 as the denominator (because you're finding the average of two items).

$$Mean = \frac{Sum\ of\ items}{Number\ of\ items} = \frac{\frac{14}{45}}{2}$$

Now, simplify the complex fraction by turning it into fraction division.

$$= \frac{14}{45} \div 2 = \frac{14}{45} \times \frac{1}{2}$$

Simplify by factoring out 2 from both the numerator and denominator; then multiply.

$$= \frac{7}{45} \times \frac{1}{1} = \frac{7}{45}$$

778. $4\frac{61}{90}$

To begin, find the sum of $3\frac{1}{3}$, $4\frac{1}{5}$, and $6\frac{1}{2}$. To do this, turn all three mixed numbers into improper fractions.

$$3\frac{1}{3} + 4\frac{1}{5} + 6\frac{1}{2} = \frac{10}{3} + \frac{21}{5} + \frac{13}{2}$$

Now, increase the terms of all three fractions to a common denominator of 30.

$$= \frac{100}{30} + \frac{126}{30} + \frac{195}{30} = \frac{421}{30}$$

Next, plug this result into the numerator of the formula for the mean, with 3 as the denominator (because you're finding the average of 3 items).

$$Mean = \frac{Sum\ of\ items}{Number\ of\ items} = \frac{\frac{421}{30}}{3}$$

Now, simplify the complex fraction by turning it into fraction division.

$$= \frac{421}{30} \div 3 = \frac{421}{30} \times \frac{1}{3} = \frac{421}{90}$$

Turn this improper fraction into a mixed number by dividing 421 by 90.

$$= 4\frac{61}{90}$$

779. $65

Kathi worked three days, averaging $60 for the three days. So she earned a total of $60 × 3 = $180 from Monday to Wednesday. She earned $40 on Monday and $75 on Tuesday, so subtract these amounts from $180 to find what she earned on Wednesday.

$$\$180 - \$40 - \$75 = \$65$$

Therefore, Kathi earned $65 on Wednesday.

780. 9 miles

Antoine hiked for 4 days at an average of 7 miles per day, so he hiked 7 × 4 = 28 miles altogether. Subtract the distances that he hiked on the first three days from 28.

$$28 - 8 - 4.5 - 6.5 = 9$$

Therefore, Antoine hiked 9 miles on the last day.

---

781. $7\frac{1}{4}$

The caterpillar crawled an average of $5\frac{1}{8}$ inches in 5 minutes, so multiply to find its total distance.

$$5\frac{1}{8} \times 5 = 25\frac{5}{8}$$

So, it traveled a total of $25\frac{5}{8}$ inches in 5 minutes. It crawled $18\frac{3}{8}$ inches in the first four minutes, so subtract to find out how far it traveled in the last minute.

$$25\frac{5}{8} - 18\frac{3}{8} = 7\frac{2}{8} = 7\frac{1}{4}$$

Therefore, it crawled $7\frac{1}{4}$ inches in the last minute.

---

782. **8 hours and 40 minutes**

Eleanor studied an average of 9 hours per day for 7 days, so she studied a total of $9 \times 7 = 63$ hours over the 7 days.

On the last day, she studied for 4 hours. On the 3 days before this, she studied for an average of 11 hours per day, so she studied for $11 \times 3 = 33$ hours. So, subtract these two values from 63.

$$63 - 4 - 33 = 26$$

Thus, she studied for a total of 26 hours on the first 3 days of the week. To find the average for these three days, divide 26 by 3.

$$\frac{26}{3} = 8\frac{2}{3}$$

Thus, she studied an average of $8\frac{2}{3}$ hours over the first three days of the week. This equals 8 hours and 40 minutes.

---

783. **17**

To calculate the weighted mean, first calculate the sum of products for the five classes.

$$(16 \times 4) + (21 \times 1)$$
$$= 64 + 21 = 85$$

Now use this result as the numerator in the formula for the mean and divide by 5.

$$Mean = \frac{Sum\ of\ items}{Number\ of\ items} = \frac{85}{5} = 17$$

Therefore, the average class size is 17 students.

---

784. **8.75 minutes**

To calculate the weighted mean, first calculate the sum of products for the eight speeches.

$$(8 \times 5) + (10 \times 3)$$
$$= 40 + 30 = 70$$

Now use this result as the numerator in the formula for the mean and divide by 8.

$$Mean = \frac{Sum\ of\ items}{Number\ of\ items} = \frac{70}{8} = 8.75$$

Therefore, the average speech length was 8.75 minutes.

785. $316

To calculate the weighted mean, first calculate the sum of products for the 10 weeks.

$$(280 \times 4) + (340 \times 6)$$
$$= 1,120 + 2,040 = 3,160$$

Now use this result as the numerator in the formula for the mean and divide by 10.

$$Mean = \frac{Sum\ of\ items}{Number\ of\ items} = \frac{3160}{10} = 316$$

Therefore, Jake's average weekly income was $316.

786. $783

To calculate the weighted mean, first calculate the sum of products for the 12 months.

$$(1,000 \times 6) + (500 \times 4) + (700 \times 2)$$
$$= 6,000 + 2,000 + 1,400 = 9,400$$

Now use this result as the numerator in the formula for the mean and divide by 12.

$$Mean = \frac{Sum\ of\ items}{Number\ of\ items} = \frac{9,400}{12} \approx 783$$

Therefore, the average savings is about $783.

787.   8 minutes and 3 seconds

Plug Angela's total time into the formula for the mean and divide by the total number of laps she ran, which was 10.

$$Mean = \frac{Sum\ of\ items}{Number\ of\ items} = \frac{31:50 + 48:40}{10}$$

Evaluate.

$$\frac{80:30}{10} = 8:03$$

Therefore, Angela's average time was 8 minutes and 3 seconds.

788.   8.5

To calculate the weighted mean, first calculate the sum of products for the 12 tests.

$$(10 \times 2) + (9 \times 5) + (8 \times 3) + (7 \times 1) + (6 \times 1)$$
$$= 20 + 45 + 24 + 7 + 6 = 102$$

Now use this result as the numerator in the formula for the mean and divide by 12.

$$Mean = \frac{Sum\ of\ items}{Number\ of\ items} = \frac{102}{12} = 8.5$$

Therefore, Kevin's average score was 8.5.

---

789. **9.4 feet**

To calculate the weighted mean, first calculate the sum of products for the 20 floors.

$$20 + (12 \times 4) + (15 \times 8)$$
$$= 20 + 48 + 120 = 188$$

Now use this result as the numerator in the formula for the mean and divide by 20.

$$Mean = \frac{Sum\ of\ items}{Number\ of\ items} = \frac{188}{20} = 9.4$$

Therefore, the average height is 9.4 feet.

---

790. **350**

To calculate the weighted mean, first calculate the sum of products for the puzzles.

$$(300 \times 2) + (1,000 \times 3) + (500 \times 4)$$
$$= 600 + 3,000 + 2,000 = 5,600$$

Now add up the number of days that she took to do these puzzles.

$$3 + 7 + 6 = 16$$

Use these results as the numerator and denominator in the formula for the mean.

$$Mean = \frac{Sum\ of\ items}{Number\ of\ items} = \frac{5,600}{16} = 350$$

Therefore, she put together an average of 350 pieces per day.

---

791. **65 mph**

To calculate the weighted mean, first calculate the sum of products for the 4 legs of the trip (be sure to convert minutes to hours).

$$(0.75 \times 75) + (1.5 \times 65) + (1.25 \times 55) + (1 \times 70)$$
$$= 56.25 + 97.5 + 68.75 + 70 = 292.5$$

Now use this result as the numerator in the formula for the mean and divide by the total time for the trip (0.75 hours + 1.5 hours + 1.25 hours + 1 hour = 4.5 hours).

$$Mean = \frac{Sum\ of\ items}{Number\ of\ items} = \frac{292.5}{4.5} = 65$$

Therefore, Gerald's average rate was 65 miles per hour.

792. **15**

The median number of any data set with an odd number of values is the middle number (when the numbers are in order). In this case, the median is 15.

793. **41**

The median number of any data set with an even number of values is the mean of the two middle numbers (when the numbers are in order). In this case, the middle numbers are 37 and 45, so find the mean as follows:

$$Mean = \frac{Sum\ of\ items}{Number\ of\ items} = \frac{37+45}{2}.$$

Simplify.

$$= \frac{82}{2} = 41$$

Therefore, the median is 41.

794. **16**

The mode of a data set is the value that occurs most frequently. In this case, 16 occurs three times, so this is the mode.

795. **0.5**

The median number of any data set with an even number of values is the mean of the two middle numbers when the numbers are listed in ascending (or descending) order. In this case, the middle numbers are 5 and 6, so find the mean as follows:

$$Mean = \frac{Sum\ of\ items}{Number\ of\ items} = \frac{5+6}{2}.$$

Simplify.

$$= \frac{11}{2} = 5.5$$

Therefore, the median is 5.5. The mode is the value that occurs most frequently in the data set, so the mode is 5. Thus, the difference between the median and the mode is $5.5 - 5 = 0.5$.

796. **13**

Calculate the mean using the formula.

$$Mean = \frac{Sum\ of\ items}{Number\ of\ items}$$
$$= \frac{1+1+11+11+11+12+13+14+14+14+63}{11}$$

Simplify:

$$= \frac{165}{11} = 15$$

Therefore, the mean is 15. The median is the middle number, which is 12. The two modes are the numbers that occur most frequently in the data set, which are 11 and 14. Therefore, 13 (the only integer between 11 and 15 that has not been ruled out) is not the mean, the median, or a mode of the data set.

---

797. 36

To calculate the number of combinations, multiply the number of possible outcomes for each die. Because there are six sides on each die, there are six different outcomes for each.

$$6 \times 6 = 36$$

---

798. 1,920

To calculate the number of combinations, multiply the number of possible outcomes for each die.

$$8 \times 12 \times 20 = 1,920$$

---

799. 56

To calculate the number of combinations, multiply the number of suits, shirts, and ties.

$$2 \times 4 \times 7 = 56$$

Therefore, Jeff had 56 possible combinations of suit, shirt, and tie.

---

800. 96

To calculate the number of combinations, multiply the number of types of eggs, meat, potatoes, and beverages.

$$4 \times 3 \times 2 \times 4 = 96$$

Therefore, 96 breakfast combinations are possible.

---

801. 1,024

There are ten questions, each of which can be answered either *yes* or *no* (two possible ways each), so calculate as follows.

$$2 \times 2 \times 2 \times 2 \times 2 \times 2 \times 2 \times 2 \times 2 \times 2 = 1,024$$

---

802. 17,576

Each of the 3 letters could be any of the 26 letters, so calculate as follows:

$$26 \times 26 \times 26 = 17,576$$

803. 1,679,616

Each of the four symbols could be any of the 10 digits or 26 letters, so there are 36 symbols in all. Calculate as follows:

$$36 \times 36 \times 36 \times 36 = 1,679,616$$

804. 24

The first letter can be any of the four possible letters (A, B, C, or D). The second can be any of the three remaining letters. The third can be either of the two remaining letters. Finally, the last letter can only be the one remaining letter tile. Multiply these four numbers together to find the total number of possible outcomes.

$$4 \times 3 \times 2 \times 1 = 24$$

Therefore, there are 24 different ways to pull four different letters from a bag.

805. 120

The first person to arrive could be any of the five people. The second person could be any of the four remaining. The third could be any of the three remaining. The fourth could be either of the two remaining. And the fifth must be the one person remaining. Calculate the total number of possible outcomes by multiplying.

$$5 \times 4 \times 3 \times 2 \times 1 = 120$$

806. 720

The first topping could be any of the six. The second could be any of the remaining five. The third could be any of the remaining four. The fourth could be any of the remaining three. The fifth could be either of the remaining two. And the sixth must be the one remaining topping. Calculate the total number of possible outcomes by multiplying.

$$6 \times 5 \times 4 \times 3 \times 2 \times 1 = 720$$

807. 40,320

The first book could be any of the eight. The second could be any of the remaining seven. The third could be any of the remaining six. The fourth could be any of the remaining five. The fifth could be any of the remaining four. The sixth could any of the remaining three. The seventh could be either of the remaining two. And the eighth must be the one remaining book. Calculate the total number of possible outcomes by multiplying.

$$8 \times 7 \times 6 \times 5 \times 4 \times 3 \times 2 \times 1 = 40,320$$

808. 6,840

The pitcher could be any of the 20 children. The catcher could be any of the remaining 19 children. And the runner could be any of the remaining 18 children. Calculate the total number of possible outcomes by multiplying.

$$20 \times 19 \times 18 = 6,840$$

809. **15,600**

The first letter could be any of the 26 letters. The second could be any of the remaining 25 letters. And the third could be any of the remaining 24 letters. Calculate the total number of possible outcomes by multiplying.

$$26 \times 25 \times 24 = 15,600$$

810. **132,600**

The first card could be any of the 52 cards. The second could be any of the remaining 51 cards. And the third could be any of the remaining 50 cards. Calculate the total number of possible outcomes by multiplying.

$$52 \times 51 \times 50 = 132,600$$

811. **43,680**

The president can be any of the 16 members. The vice-president can be any of the remaining 15 members. The treasurer can be any of the remaining 14 members. And the secretary can be any of the remaining 13 members. Calculate the total number of possible outcomes by multiplying.

$$16 \times 15 \times 14 \times 13 = 43,680$$

812. **27,216**

The first digit can be any of the nine digits, 1 through 9. The second digit can be any of the nine remaining digits from 0 through 9. The third can be any of the eight remaining digits. The fourth can be any of the seven remaining digits. And the fifth can be any of the six remaining digits. Calculate the total number of possible outcomes by multiplying.

$$9 \times 9 \times 8 \times 7 \times 6 = 27,216$$

813. **2,160**

The first letter must be one of the three vowels. The second can be any of the remaining six letters. The third can be any of the remaining five letters. The fourth can be any of the remaining four letters. The fifth letter can be any of the remaining three letters. The sixth letter can be either of the remaining two letters. And the seventh letter must be the one remaining letter. Calculate the total number of possible outcomes by multiplying.

$$3 \times 6 \times 5 \times 4 \times 3 \times 2 \times 1 = 2,160$$

814. **432**

The first letter must be one of the three vowels. The second letter must be one of the remaining two vowels. The third letter must be one of the four consonants. The fourth letter must be one of the remaining three consonants. The fifth letter can be any of the remaining three letters. The sixth letter can be either of the remaining two letters. And the seventh letter must be the one remaining letter. Calculate the total number of possible outcomes by multiplying.

$$3 \times 2 \times 4 \times 3 \times 3 \times 2 \times 1 = 432$$

815.   36

The first arrival was one of the three women, the second was one of the two remaining women, and the third was the one remaining woman. Then, the fourth arrival was one of the three men, the fifth was either of the two remaining men, and the sixth was the one remaining man. Multiply these six numbers together to calculate the total number of possible outcomes.

$$3 \times 2 \times 1 \times 3 \times 2 \times 1 = 36$$

816.   36

Each man arrived just after a woman, so the women arrived first, third, and fifth, and the men arrived second, fourth, and sixth. The first arrival was one of the three women, the second was one of the three men, the third was one of the two remaining women, the fourth was one of the two remaining men, the fifth arrival was the one remaining woman, and the sixth was the one remaining man. Multiply these six numbers together to calculate the total number of possible outcomes.

$$3 \times 3 \times 2 \times 2 \times 1 \times 1 = 36$$

817. $\frac{1}{10}$

When you pull one ticket from a bag containing ten tickets, there are a total of ten possible outcomes. In this case, there is only one target outcome: pulling the ticket with the number 1. Plug this information into the formula for probability.

$$Probability = \frac{Target\ outcomes}{Total\ outcomes} = \frac{1}{10}$$

Therefore, the probability is $\frac{1}{10}$.

818. $\frac{1}{2}$

When you pull one ticket from a bag containing ten tickets, there are a total of ten possible outcomes. In this case, there are five target outcomes: pulling the tickets with the numbers 2, 4, 6, 8, or 10. Plug this information into the formula for probability.

$$Probability = \frac{Target\ outcomes}{Total\ outcomes} = \frac{5}{10} = \frac{1}{2}$$

Therefore, the probability is $\frac{1}{2}$.

819. $\frac{2}{5}$

When you pull one ticket from a bag containing ten tickets, there are a total of ten possible outcomes. In this case, there are four target outcomes: pulling the tickets with the numbers 7, 8, 9, or 10. Plug this information into the formula for probability.

$$Probability = \frac{Target\ outcomes}{Total\ outcomes} = \frac{4}{10} = \frac{2}{5}$$

Therefore, the probability is $\frac{2}{5}$.

820. $\frac{2}{9}$

When you pull one ticket from a bag containing ten tickets, there are a total of ten possible outcomes. Then, when you pull a second ticket from the bag, there are nine possible outcomes. Therefore, there are a total of $10 \times 9 = 90$ possible outcomes.

In this case, there are five target outcomes for the first pull (the numbers 1, 3, 5, 7, or 9) and four target outcomes for the second pull (any of four odd numbers that remain after the first pull. Therefore, there are $5 \times 4 = 20$ target outcomes.

Plug this information into the formula for probability.

$$Probability = \frac{Target\ outcomes}{Total\ outcomes} = \frac{20}{90} = \frac{2}{9}$$

Therefore, the probability is $\frac{2}{9}$.

821. $\frac{1}{6}$

When you roll a six-sided die, there are a total of six possible outcomes. In this case, there is only one target outcome: rolling the number 2. Plug this information into the formula for probability.

$$Probability = \frac{Target\ outcomes}{Total\ outcomes} = \frac{1}{6}$$

Therefore, the probability is $\frac{1}{6}$.

822. $\frac{2}{3}$

When you roll a six-sided die, there are a total of six possible outcomes. In this case, there are four target outcomes: rolling the numbers 3, 4, 5, and 6. Plug this information into the formula for probability.

$$Probability = \frac{Target\ outcomes}{Total\ outcomes} = \frac{4}{6} = \frac{2}{3}$$

Therefore, the probability is $\frac{2}{3}$.

823. $\frac{5}{6}$

When you roll a six-sided die, there are a total of six possible outcomes. In this case, there are five target outcomes: rolling the numbers 1, 3, 4, 5, and 6. Plug this information into the formula for probability.

$$Probability = \frac{Target\ outcomes}{Total\ outcomes} = \frac{5}{6}$$

Therefore, the probability is $\frac{5}{6}$.

824. $\frac{1}{36}$

When you roll two six-sided dice, there are a total of six possible outcomes for the first die and six possible outcomes for the second die. Thus, the total number of outcomes is $6 \times 6 = 36$.

In this case, there is one target outcome: rolling 6 on the first die and 6 on the second.

Plug this information into the formula for probability.

$$Probability = \frac{Target\ outcomes}{Total\ outcomes} = \frac{1}{36}$$

Therefore, the probability is $\frac{1}{36}$.

825. $\frac{1}{12}$

When you roll two six-sided dice, there are a total of six possible outcomes for the first die and six possible outcomes for the second die. Thus, the total number of outcomes is $6 \times 6 = 36$.

In this case, there are three target outcomes: rolling 4 on the first die and 6 on the second, rolling 5 on the first die and 5 on the second, and rolling 6 on the first die and 4 on the second.

Plug this information into the formula for probability.

$$Probability = \frac{Target\ outcomes}{Total\ outcomes} = \frac{3}{36} = \frac{1}{12}$$

Therefore, the probability is $\frac{1}{12}$.

826. $\frac{2}{9}$

When you roll two six-sided dice, there are a total of six possible outcomes for the first die and six possible outcomes for the second die. Thus, the total number of outcomes is $6 \times 6 = 36$.

To count the number of target outcomes, first count the number of 11s and then the number of 7s.

There are two target outcomes that add up to 11: rolling 5 on the first die and 6 on the second, and rolling 6 on the first die and 5 on the second.

There are six target outcomes that add up to 7: rolling 1 on the first die and 6 on the second, rolling 2 on the first die and 5 on the second, rolling 3 on the first die and 4 on the second, rolling 4 on the first die and 3 on the second, rolling 5 on the first die and 2 on the second, and rolling 6 on the first die and 1 on the second.

Therefore, there are 8 target outcomes and 36 total outcomes. Plug this information into the formula for probability.

$$Probability = \frac{Target\ outcomes}{Total\ outcomes} = \frac{8}{36} = \frac{2}{9}$$

Therefore, the probability is $\frac{2}{9}$.

---

827. $\frac{1}{36}$

When you roll three six-sided dice, there are a total of six possible outcomes for the first die, six possible outcomes for the second die, and six possible outcomes for the third die. Thus, the total number of outcomes is $6 \times 6 \times 6 = 216$.

There are six target outcomes that add up to 16:

6 + 6 + 4

6 + 4 + 6

6 + 5 + 5

5 + 6 + 5

5 + 5 + 6

4 + 6 + 6

Therefore, there are 6 target outcomes and 216 total outcomes. Plug this information into the formula for probability.

$$Probability = \frac{Target\ outcomes}{Total\ outcomes} = \frac{6}{216} = \frac{1}{36}$$

Therefore, the probability is $\frac{1}{36}$.

---

828. $\frac{1}{13}$

When you pick a card from a deck of 52 cards, there are a total of 52 possible outcomes. In this case, there are four target outcomes: pulling one of the four aces. Plug this information into the formula for probability.

$$Probability = \frac{Target\ outcomes}{Total\ outcomes} = \frac{4}{52} = \frac{1}{13}$$

Therefore, the probability is $\frac{1}{13}$.

---

829. $\frac{1}{4}$

When you pick a card from a deck of 52 cards, there are a total of 52 possible outcomes. In this case, there are 13 target outcomes: picking one of the 13 hearts. Plug this information into the formula for probability.

$$Probability = \frac{Target\ outcomes}{Total\ outcomes} = \frac{13}{52} = \frac{1}{4}$$

Therefore, the probability is $\frac{1}{4}$.

830. $\dfrac{3}{13}$

When you pick a card from a deck of 52 cards, there are a total of 52 possible outcomes. In this case, there are 12 target outcomes: picking one of the four kings, one of the four queens, or one of the four jacks. Plug this information into the formula for probability.

$$\text{Probability} = \frac{Target\ outcomes}{Total\ outcomes} = \frac{12}{52} = \frac{3}{13}$$

Therefore, the probability is $\dfrac{3}{13}$.

831. $\dfrac{1}{221}$

When you pick 2 cards from a deck of 52 cards, there are a total of 52 possible outcomes for the first card and 51 possible outcomes for the second card. Thus, $52 \times 51 = 2,652$ total outcomes are possible.

In this case, there are four target outcomes for the first card (picking one of the four aces) and three target outcomes for the second card (picking one of the three remaining aces). Thus, $4 \times 3 = 12$ target outcomes are possible.

Plug this information into the formula for probability.

$$\text{Probability} = \frac{Target\ outcomes}{Total\ outcomes} = \frac{12}{2,652} = \frac{1}{221}$$

Therefore, the probability is $\dfrac{1}{221}$.

832. $\dfrac{1}{270,725}$

When you pick 4 cards from a deck of 52 cards, there are a total of 52 possible outcomes for the first card, 51 for the second card, 50 for the third card, and 49 for the fourth card. Thus, $52 \times 51 \times 50 \times 49 = 6,497,400$ total outcomes are possible.

In this case, there are four target outcomes for the first card (picking one of the four aces), three for the second card (picking one of the three remaining aces), two for the third (picking one of the two remaining aces), and one for the fourth (picking the one remaining ace). Thus, $4 \times 3 \times 2 \times 1 = 24$ target outcomes are possible.

Plug this information into the formula for probability.

$$\text{Probability} = \frac{Target\ outcomes}{Total\ outcomes} = \frac{24}{6,497,400} = \frac{1}{270,725}$$

Therefore, the probability is $\dfrac{1}{270,725}$

833. $\dfrac{1}{2}$

The first person to arrive was one of six people, so the total number of possible outcomes was six. Of these, there are three target outcomes (each of the three women arriving first). Plug this information into the formula for probability.

$$\text{Probability} = \frac{Target\ outcomes}{Total\ outcomes} = \frac{3}{6} = \frac{1}{2}$$

Therefore, the probability is $\dfrac{1}{2}$.

834. $\dfrac{1}{20}$

There are a total of six possible outcomes for the first person, five possible outcomes for the second person, and four possible outcomes for the third person. Thus, the total number of outcomes is $6 \times 5 \times 4 = 120$.

There are three possible target outcomes for the first person (one of the three women arrives first), two for the second person (one of the remaining two women arrives second) and one for the third (the one remaining woman arrives third). Thus, the number of target outcomes is $3 \times 2 \times 1 = 6$.

Therefore, there are 6 target outcomes and 120 total outcomes. Plug this information into the formula for probability.

$$Probability = \frac{Target\ outcomes}{Total\ outcomes} = \frac{6}{120} = \frac{1}{20}$$

Therefore, the probability is $\dfrac{1}{20}$.

835. $\dfrac{1}{20}$

There are a total of six possible outcomes for the first person, five possible outcomes for the second person, four for the third person, three for the fourth person, two for the fifth person, and one for the sixth person. Thus, the total number of outcomes is $6 \times 5 \times 4 \times 3 \times 2 \times 1 = 120$.

Each man arrived just after a woman, so the women arrived first, third, and fifth, and the men arrived second, fourth, and sixth. The first arrival was one of the three women, the second was one of the three men, the third was one of the two remaining women, the fourth was one of the two remaining men, the fifth arrival was the one remaining woman, and the sixth was the one remaining man. Multiply these six numbers together:

$$3 \times 3 \times 2 \times 2 \times 1 \times 1 = 36$$

So the number of total outcomes is 720, and the number of target outcomes is 36. Plug these numbers into the formula for probability.

$$Probability = \frac{Target\ outcomes}{Total\ outcomes} = \frac{36}{120} = \frac{1}{20}$$

Therefore, the probability is $\dfrac{1}{20}$.

836. {1, 3, 5, 6, 7, 8, 9}

$P \bigcup Q$ is the union of $P = \{1, 3, 5, 7, 9\}$ and $Q = \{6, 7, 8\}$. The union includes every element in *either* set.

837. {7}

$P \bigcap Q$ is the intersection of $P = \{1, 3, 5, 7, 9\}$ and $Q = \{6, 7, 8\}$. The intersection includes every element in *both* sets.

838. {1, 3, 5, 9}

$P - Q$ is the relative complement of $P = \{1, 3, 5, 7, 9\}$ and $Q = \{6, 7, 8\}$. The relative complement includes only elements of the first set (P) that are *not* in the second set (Q).

839. {6, 8}

$Q - P$ is the relative complement of $Q = \{6, 7, 8\}$ and $P = \{1, 3, 5, 7, 9\}$. The relative complement includes only elements of the first set (Q) that are *not* in the second set (P).

840. {3, 6, 7, 8, 9}

$Q \cup S$ is the union of $Q = \{6, 7, 8\}$ and $S = \{3, 6, 9\}$, which includes every element in *either* set.

841. $\varnothing$

$R \cap S$ is the intersection of $R = \{1, 2, 4, 5\}$ and $S = \{3, 6, 9\}$, which includes every element in *both* sets. The two sets have no element in common, so the intersection of these sets is the empty set.

842. {1, 5}

Begin by finding $P \cup Q$. This is the union of $P = \{1, 3, 5, 7, 9\}$ and $Q = \{6, 7, 8\}$. The union includes every element in *either* set, so

$P \cup Q = \{1, 3, 5, 6, 7, 8, 9\}$

Now, find the intersection of this set and $R = \{1, 2, 4, 5\}$. The intersection includes every element in *both* sets, so

$(P \cup Q) \cap R = \{1, 5\}$

843. {1, 3, 5, 7, 9}

Begin by finding $Q \cap R$. This is the intersection of $Q = \{6, 7, 8\}$ and $R = \{1, 2, 4, 5\}$. The intersection includes every element in *both* sets, so

$Q \cap R = \varnothing$

Now, find the union of the empty set and $P = \{1, 3, 5, 7, 9\}$. The union includes every element in *either* set, so

$P \cup (Q \cap R) = \{1, 3, 5, 7, 9\}$

844. {1, 5}

Begin by finding $Q \cup S$. This is the union of $Q = \{6, 7, 8\}$ and $S = \{3, 6, 9\}$, which includes every element in *either* set, so

$Q \cup S = \{3, 6, 7, 8, 9\}$

Now, find the relative complement of $P = \{1, 3, 5, 7, 9\}$ and this set — that is, the elements of P that are *not* in $Q \cup S$:

$$P - (Q \cup S) = \{1, 5\}$$

845. **{1, 3, 5, 6, 9}**

Begin by finding $P - Q$. This is the relative complement of $P = \{1, 3, 5, 7, 9\}$ and $Q = \{6, 7, 8\}$. The relative complement includes only elements of the first set (P) that are *not* in the second set (Q), so

$$P - Q = \{1, 3, 5, 9\}$$

Now, find the union of this set and $S = \{3, 6, 9\}$. The union includes every element in *either* set, so

$$(P - Q) \cup S = \{1, 3, 5, 6, 9\}$$

846. $\varnothing$

Begin by finding $Q - S$. This is the relative complement of $Q = \{6, 7, 8\}$ and $S = \{3, 6, 9\}$. The relative complement includes only elements of the first set (Q) that are *not* in the second set (S), so

$$Q - S = \{7, 8\}$$

Now, find the intersection of this set and $R = \{1, 2, 4, 5\}$. The intersection includes every element in *both* sets, so

$$(Q - S) \cap R = \varnothing$$

847. **{1, 5, 7}**

Begin by finding $Q \cup R$ and $P - S$:

$$Q \cup R = \{1, 2, 4, 5, 6, 7, 8\}$$
$$P - S = \{1, 5, 7\}$$

Now, find the intersection of these two sets:

$$(Q \cup R) \cap (P - S) = \{1, 5, 7\}$$

848. **{..., −4, −2, 0, 2, 4, ...}**

The set of integers is $\{..., -2, -1, 0, 1, 2, ...\}$, and the set of even integers is $\{..., -4, -2, 0, 2, 4, ...\}$. The intersection includes every element in *both* sets.

849. **{1, 3, 5, 7, ...}**

The set of positive integers is $\{1, 2, 3, 4, ...\}$, and the set of even integers is $\{..., -4, -2, 0, 2, 4, ...\}$. The relative complement includes only elements of the first set that are *not* in the second set.

850.   {..., –3, –1, 2, 4, 6, ...}

The set of odd negative integers is {..., –7, –5, –3, –1}, and the set of even positive integers is {2, 4, 6, 8, ...}. The union includes elements that are in *either* set.

851.   ∅

The set of odd negative integers is {..., –7, –5, –3, –1}, and the set of even positive integers is {2, 4, 6, 8, ...}. The intersection includes all elements that are in *both* sets.

852.   {..., 3, 4, 5, 6, 7}

The complement of a set includes every element that is in the universal set but *not* in the set itself. The universal set in this case is {..., –2, –1, 0, 1, 2, ...}, and the set of integers greater than 7 is {8, 9, 10, 11, 12, ...}. Therefore, the complement of this set is all the integers less than or equal to 7, {..., 3, 4, 5, 6, 7}.

853.   {..., –4, –2, 0, 2, 4, ...}

The complement of a set includes every element that is in the universal set but *not* in the set itself. The universal set in this case is {..., –2, –1, 0, 1, 2, ...}, and the set of odd integers is {..., –3, –1, 1, 3, 5, ...}. Thus, the complement of the set of odd integers is the set of even integers {..., –4, –2, 0, 2, 4, ...}.

854.   {..., –2, –1, 0, 1, 2, ...}

The complement of a set includes every element that is in the universal set but *not* in the set itself. The universal set in this case is {..., –2, –1, 0, 1, 2, ...}, and the set itself is the empty set, ∅. Because the empty set has no elements, no elements need to be removed from the universal set to form its complement. Therefore, the complement of ∅ is {..., –2, –1, 0, 1, 2, ...}.

855.   {..., –5, –4, –3, –2, –1}

The complement of a set includes every element that is in the universal set but *not* in the set itself. The universal set in this case is {..., –2, –1, 0, 1, 2, ...}, and the set of non-negative integers is {0, 1, 2, 3, 4 ,...}. Therefore, the complement of this set is {..., –5, –4, –3, –2, –1}.

Another way to think about it: The complement of the set of *non-negative* integers is the set of *negative* integers.

856.   29

The diagram shows that 6 students are seniors only, 3 are honors students only, 8 are both seniors and honors students, and 12 are neither. Thus, the club has 6 + 3 + 8 + 12 = 29 members.

857. 27

The diagram shows that 20 people are surnamed Kinney only, 9 live out of state only, and 6 are neither surnamed Kinney nor live out of state. This accounts for 20 + 9 + 6 = 35 of the 42 attendees. Thus, 7 people both are surnamed Kinney and live out of state. So, a total of 20 + 7 = 27 people are surnamed Kinney.

858. 2

The diagram shows that 10 people were in _12 Angry Men_ but not in _Long Day's Journey Into Night_. And _12 Angry Men_ had 13 people, so 3 were in both plays. _Long Day's Journey Into Night_ had 5 people, so 2 were in this play but not _12 Angry Men_.

859. 3

The board includes 2 officers who have served more than one term. Thus, of the 7 officers, the other 5 are serving their first term. And of the 10 people who have served more than one term, 8 are nonofficers. This accounts for 15 board members, so the remaining 3 are nonofficers who are serving their first term. The following Venn diagram shows this information:

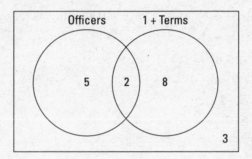

860. 11

Three of the children own neither a cat nor a dog, so 21 students own at least one of these animals. Of these, 15 own a cat and 10 own a dog. 15 + 10 = 25, which is 4 greater than 21. Therefore, exactly 4 students own both a cat and a dog.

You can see this breakdown in the following Venn diagram:

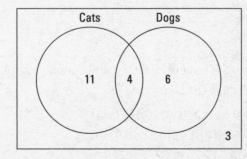

So, of the 15 students who own at least one cat, 11 own at least one cat but no dog.

861. 7

Substitute 9 for x and 4 for y and simplify as follows:

$3x - 5y$

$= 3(9) - 5(4)$

$= 27 - 20$

$= 7$

862. 51

Substitute 5 for x and –2 for y and simplify as follows:

$x^2 - 8y + 10$

$= 5^2 - 8(-2) + 10$

$= 25 + 16 + 10$

$= 51$

863. 83

Substitute –6 for x and –1 for y and simplify as follows:

$-2x^2 y + 11$

$= -2(-6)^2(-1) + 11$

$= -2(-6)(-6)(-1) + 11$

$= 72 + 11$

$= 83$

864. –65

Substitute –2 for x and 3 for y and simplify as follows:

$4x^3 + 5xy - y$

$= 4(-2)^3 + 5(-2)(3) - 3$

$= 4(-2)(-2)(-2) + 5(-2)(3) - 3$

$= -32 - 30 - 3$

$= -65$

865. –0.75

Substitute 0.5 for x and –0.5 for y and simplify as follows:

$3x^2 + 5xy - 0.25$

$= 3(0.5)^2 + 5(0.5)(-0.5) - 0.25$

$= 3(0.5)(0.5) + 5(0.5)(-0.5) - 0.25$

$= 0.75 - 1.25 - 0.25$

$= -0.75$

866. **0.1805**

Substitute 0.1 for x and 3 for y and simplify as follows:

$$5\left(x^2y^2+x\right)^2$$
$$=5\left[\left(0.1^2\right)\left(3^2\right)+0.1\right]^2$$
$$=5\left[(0.01)(9)+0.1\right]^2$$
$$=5\left[0.09+0.1\right]^2$$
$$=5\left[0.19\right]^2$$
$$=5\left[0.0361\right]$$
$$=0.1805$$

867. **–11.62**

Substitute 7 for x and 9 for y and simplify as follows:

$$0.7\left(0.1xy+0.2x-0.3y^2\right)$$
$$=0.7\left[0.1(7)(9)+0.2(7)-0.3(9)^2\right]$$
$$=0.7\left[0.1(7)(9)+0.2(7)-0.3(9)(9)\right]$$
$$=0.7\left[6.3+1.4-24.3\right]$$
$$=0.7\left[-16.6\right]$$
$$=-11.62$$

868. $-\dfrac{73}{40}$

Substitute 5 for x and –8 for y and simplify as follows:

$$\frac{x}{y}+\frac{3y}{4x}$$
$$=\frac{5}{-8}+\frac{3(-8)}{4(5)}$$
$$=-\frac{5}{8}-\frac{24}{20}$$
$$=-\frac{25}{40}-\frac{48}{40}$$
$$=-\frac{73}{40}$$

869. −1,458

Substitute −1 for x and 2 for y and simplify as follows:

$$\frac{(10xy+y)^3}{2y}$$

$$=\frac{\left[10(-1)(2)+2\right]^3}{2(2)}$$

$$=\frac{\left[-20+2\right]^3}{4}$$

$$=\frac{\left[-18\right]^3}{4}$$

$$=\frac{-5,832}{4}$$

$$=-1,458$$

870. $\frac{16}{225}$

Substitute −2 for x and 3 for y and simplify as follows:

$$\left(\frac{x}{y}\right)^4\left(\frac{2y}{5x}\right)^2$$

$$=\left(\frac{-2}{3}\right)^4\left(\frac{2(3)}{5(-2)}\right)^2$$

$$=\left(\frac{-2}{3}\right)^4+\left(\frac{6}{-10}\right)^2$$

$$=\left(\frac{-2}{3}\right)^4\left(\frac{3}{-5}\right)^2$$

$$=\left(\frac{16}{81}\right)\left(\frac{9}{25}\right)$$

$$=\left(\frac{16}{9}\right)\left(\frac{1}{25}\right)$$

$$=\frac{16}{225}$$

871. $7x+2y$

Simplify by combining the two x terms and the two y terms.

$$2x+3y+5x-y$$

$$=(2x+5x)+(3y-y)$$

$$=7x+2y$$

872. $8x^3 - 7x^2 + 3x - 18$

Simplify by combining each pair of like terms.
$$2x^3 + 3x^2 + 5x - 9 + 6x^3 - 10x^2 - 2x - 9$$
$$= \left(2x^3 + 6x^3\right) + \left(3x^2 - 10x^2\right) + \left(5x - 2x\right) + \left(-9 - 9\right)$$
$$= 8x^3 - 7x^2 + 3x - 18$$

873. $7.9x + 5y + 4xy + 4.8$

Simplify by combining the two x terms, the three constant terms, and the two xy terms.
$$8x + 0.1 + 5y + 3xy + 5 - 0.1x - 0.3 + xy$$
$$= \left(8x - 0.1x\right) + \left(5y\right) + \left(3xy + xy\right) + \left(0.1 + 5 - 0.3\right)$$
$$= 7.9x + 5y + 4xy + 4.8$$

874. $\frac{3}{5}x + \frac{1}{3}x^2 + \frac{3}{8}x^3$

Simplify by combining the two x terms and the two x^3 terms.
$$x + \frac{1}{3}x^2 - \frac{1}{4}x^3 - \frac{2}{5}x + \frac{5}{8}x^3$$
$$= \left(x - \frac{2}{5}x\right) + \frac{1}{3}x^2 + \left(-\frac{1}{4}x^3 + \frac{5}{8}x^3\right)$$
$$= \frac{3}{5}x + \frac{1}{3}x^2 + \frac{3}{8}x^3$$

875. $20x^7$

Multiply the coefficients ($4 \cdot 5 = 20$); then multiply the x variables by adding the exponents ($3 + 4 = 7$).
$$\left(4x^3\right)\left(5x^4\right)$$
$$= 20\left(x^3\right)\left(x^4\right)$$
$$= 20x^7$$

876. $12x^3y^6$

Multiply the coefficients ($2 \cdot 6 = 12$); then multiply the like variables by adding the exponents of the x variables ($2 + 1 = 3$) and y variables ($2 + 4 = 6$).
$$\left(2x^2y^2\right)\left(6xy^4\right)$$
$$= 12\left(x^2y^2\right)\left(xy^4\right)$$
$$= 12x^3\left(y^2\right)\left(y^4\right)$$
$$= 12x^3y^6$$

877. $21x^5y^3z^5$

Multiply the coefficients $(7 \cdot 1 \cdot 3 = 21)$; then multiply the like variables by adding the exponents of the x variables $(2 + 3 = 5)$, y variables $(1 + 1 + 1 = 3)$, and z variables $(1 + 4 = 5)$.

$$\left(7x^2yz\right)\left(x^3y\right)\left(3yz^4\right)$$
$$= 21\left(x^2yz\right)\left(x^3y\right)\left(yz^4\right)$$
$$= 21x^5\left(yz\right)\left(y\right)\left(yz^4\right)$$
$$= 21x^5y^3\left(z\right)\left(z^4\right)$$
$$= 21x^5y^3z^5$$

878. $81x^6$

Apply the rule for simplifying exponents: Take the coefficient (9) to the power of 2, and multiply the exponent of x (3) by 2. To show why this works, I do this in two steps:

$$\left(9x^3\right)^2$$
$$= \left(9x^3\right)\left(9x^3\right)$$
$$= 81x^6$$

879. $216x^6y^{12}z^{15}$

Apply the rule for simplifying exponents: Take the coefficient (6) to the power of 3, and multiply the exponents of x, y, and z by 3:

$$\left(6x^2y^4z^5\right)^3 = 216x^6y^{12}z^{15}$$

880. $768x^7y^{14}$

Begin by expanding the exponents.

$$\left(4xy\right)^2\left(6x^2\right)\left(2y^4x\right)^3$$
$$= \left(16x^2y^2\right)\left(4xy\right)\left(6x^2\right)\left(8y^{12}x^3\right)$$

Now, multiply the coefficients, and then add the exponents of the x variables and the y variables.

$$= 768x^7y^{14}$$

881. $2x^3y$

Cancel the common factor of the coefficients (2) in both the numerator and denominator.

$$\frac{8x^4y^3}{4xy^2}$$

$$= \frac{2x^4y^3}{xy^2}$$

Next, simplify the variables by subtracting the exponents in the numerator minus the corresponding exponents in the denominator.

$$= 2x^3y$$

882. $2x^4$

Begin by applying the rule for simplifying exponents in both the numerator and denominator:

$$\frac{\left(4x^5\right)^2}{\left(2x^2\right)^3}$$

$$= \frac{16x^{10}}{8x^6}$$

Now, cancel the common factor of the coefficients (8) in both the numerator and denominator. Then simplify the variables by subtracting the exponents in the numerator minus the exponents in the denominator.

$$= \frac{16x^{10}}{8x^6}$$

$$= \frac{2x^{10}}{x^6}$$

$$= 2x^4$$

883. y

Begin by applying the rule for simplifying exponents in both the numerator and denominator; then simplify.

$$\frac{x\left(4xy^2\right)^3}{y^5\left(8x^2\right)^2}$$

$$= \frac{x\left(64x^3y^6\right)}{y^5\left(64x^4\right)}$$

$$= \frac{64x^4y^6}{64x^4y^5}$$

Now, cancel the common factor of the coefficients (64) in both the numerator and denominator. Then simplify the variables by subtracting the exponents in the numerator minus the corresponding exponents in the denominator.

$$= \frac{x^4 y^6}{x^4 y^5}$$

$$= \frac{y^6}{y^5}$$

$$= y$$

884. $3y + 3$

To simplify, first remove the parentheses; then combine like terms.

$$x + (3y - x - 5) + 8$$
$$= x + 3y - x - 5 + 8$$
$$= (x - x) + 3y + (-5 + 8)$$
$$= 3y + (-5 + 8)$$
$$= 3y + 3$$

885. $-5x - y + 3$

To simplify, first negate all the terms inside the first set of parentheses; then remove both sets of parentheses and combine like terms.

$$3y - (6x + 4y - 5) + (x - 2)$$
$$= 3y - 6x - 4y + 5 + x - 2$$
$$= (-6x + x) + (3y - 4y) + (5 - 2)$$
$$= -5x - y + 3$$

886. $18x^2 - 24x + 3xy + 11$

To simplify, first distribute $3x$ among all terms inside the first set of parentheses and negate all terms inside the second set of parentheses; then remove both sets of parentheses.

$$3x(6x + 4y - 8) - (9xy - 11)$$
$$= 18x^2 + 12xy - 24x - 9xy + 11$$

Now, combine like terms.

$$= 18x^2 - 24x + (12xy - 9xy) + 11$$
$$= 18x^2 - 24x + 3xy + 11$$

887. $-42xy - 9xyz + 4yz$

To simplify, first distribute $-6xy$ among all terms inside the first set of parentheses and $-yz$ among all terms inside the second set of parentheses; then remove both sets of parentheses.

$$-6xy(7+z) - yz(3x-4) = -42xy - 6xyz - 3xyz + 4yz$$

Now, combine like terms.

$$= -42xy - 9xyz + 4yz$$

888. $6x^3 - 2x^2 + 6x + 63$

To simplify, distribute to remove all three sets of parentheses.

$$x^2(6x+4) + 3x(x+2) - 9(x^2-7)$$
$$= 6x^3 + 4x^2 + 3x(x+2) - 9(x^2-7)$$
$$= 6x^3 + 4x^2 + 3x^2 + 6x - 9(x^2-7)$$
$$= 6x^3 + 4x^2 + 3x^2 + 6x - 9x^2 + 63$$

Simplify by combining like terms.

$$= 6x^3 - 2x^2 + 6x + 63$$

889. $x^2 - x - 12$

Multiply the two expressions by using the "FOIL" method.

$$(x+3)(x-4) = x^2 - 4x + 3x - 12$$

Simplify by combining like terms.

$$= x^2 - x - 12$$

890. $10x^2 + 11x + 3$

Multiply the two expressions by using the "FOIL" method.

$$(2x+1)(5x+3) = 10x^2 + 6x + 5x + 3$$

Simplify by combining like terms.

$$= 10x^2 + 11x + 3$$

891. $x^3 + 7x^2 - 2x - 14$

Multiply the two expressions by using the "FOIL" method.

$$(x^2-2)(x+7) = x^3 + 7x^2 - 2x - 14$$

892. $4x^3 - 64x^2 + 240x$

Begin by distributing $4x$ over $(x-6)$.
$$4x(x-6)(x-10) = (4x^2 - 24x)(x-10)$$

Multiply the two resulting expressions by using the "FOIL" method.
$$= 4x^3 - 40x^2 - 24x^2 + 240x$$

Simplify by combining like terms:
$$= 4x^3 - 64x^2 + 240x$$

893. $x^3 + 3x^2 - 6x - 8$

Multiply the first two expressions by using the "FOIL" method.
$$(x+1)(x-2)(x+4) = (x^2 - 2x + x - 2)(x+4)$$

Simplify the first expression by combining like terms.
$$= (x^2 - x - 2)(x+4)$$

Now, multiply each term in the first expression by each term in the second expression.
$$= x^3 + 4x^2 - x^2 - 4x - 2x - 8$$

Simplify by combining like terms.
$$= x^3 + 3x^2 - 6x - 8$$

894. $x(x-3)$

You can factor an x out of both terms.
$$x^2 - 3x = x(x-3)$$

895. $x^2(x^3 + 1)$

You can factor an x^2 out of both terms.
$$x^5 + x^2 = x^2(x^3 + 1)$$

896. $x^6(6x^2 - x - 4)$

You can factor an x^6 out of all three terms.
$$6x^8 - x^7 - 4x^6 = x^6(6x^2 - x - 4)$$

897. $2x^3\left(-6x^6+3x^3+2\right)$

The greatest common factor of 12, 6, and 4 is 2. And the greatest common factor of the x variables has the smallest exponent among the three terms, 3. Thus, you can factor a $2x^3$ out of all three terms.

$$-12x^9+6x^6+4x^3=2x^3\left(-6x^6+3x^3+2\right)$$

898. $3x^4\left(8x^6+5x^5+3\right)$

The greatest common factor of 24, 15, and 9 is 3. And the greatest common factor of the x variables has the smallest exponent among the three terms, 4. Thus, you can factor a $3x^4$ out of all three terms.

$$24x^{10}+15x^9+9x^4=3x^4\left(8x^6+5x^5+3\right)$$

899. $x^2y\left(y^2+x^5y^6+x^2\right)$

The greatest common factor of the x variables has the smallest exponent among the three terms, 2, and the greatest common factor of the y variables has an exponent of 1. Thus, you can factor an x^2y out of all three terms.

$$x^2y^3+x^7y^7+x^4y=x^2y\left(y^2+x^5y^6+x^2\right)$$

900. $4x^6y^8\left(2x^5y^6+5x^3-10y^2\right)$

The greatest common factor of 8, 20, and 40 is 4. The greatest common factor of the x variables has the smallest exponent among the three terms, 6. And the greatest common factor of the y variables has the smallest exponent of all three terms, 8. Thus, you can factor a $4x^6y^8$ out of all three terms.

$$8x^{11}y^{14}+20x^9y^8-40x^6y^{10}=4x^6y^8\left(2x^5y^6+5x^3-10y^2\right)$$

901. $6xyz^3\left(6z^2-4y^7z+15x^5y^3\right)$

The greatest common factor of 36, 24, and 90 is 6. And the greatest common x, y, and z exponents are, respectively, 1, 1, and 3. Thus, you can factor a $6xyz^3$ out of all three terms.

$$36xyz^5-24xy^8z^4+90x^6y^4z^3=6xyz^3\left(6z^2-4y^7z+15x^5y^3\right)$$

902. $(x+8)(x-8)$

Both terms are perfect squares, so you can use the rule for factoring the difference of two squares.

$$x^2-64=(x+8)(x-8)$$

903. $(3x+2)(3x-2)$

Both terms are perfect squares, so you can use the rule for factoring the difference of two squares.

$$9x^2 - 4 = (3x+2)(3x-2)$$

904. $(7x+10y)(7x-10y)$

Both terms are perfect squares, so you can use the rule for factoring the difference of two squares.

$$49x^2 - 100y^2 = (7x+10y)(7x-10y)$$

905. $(x^2+4y^5)(x^2-4y^5)$

Both terms are perfect squares, so you can use the rule for factoring the difference of two squares.

$$x^4 - 16y^{10} = (x^2+4y^5)(x^2-4y^5)$$

906. $(x+2)(x+7)$

Begin by generating a list of all possible factor pairs of integers (both negative and positive) that multiply to 14 (the constant).

$$(1)(14) = 14$$
$$(-1)(-14) = 14$$
$$(2)(7) = 14$$
$$(-2)(-7) = 14$$

Identify the factor pair whose sum is 9 (the coefficient of the x term).

$$1+14 = 15$$
$$-1+(-14) = -15$$
$$2+7 = 9$$
$$-2+(-7) = -9$$

So, factor using the numbers 2 and 7, as follows:

$$x^2 + 9x + 14 = (x+2)(x+7)$$

907. $(x-2)(x-9)$

Begin by generating a list of all possible factor pairs of integers (both negative and positive) that multiply to 18 (the constant).

$$(1)(18) = 18$$
$$(-1)(-18) = 18$$
$$(2)(9) = 18$$
$$(-2)(-9) = 18$$
$$(3)(6) = 18$$
$$(-3)(-6) = 18$$

Identify the factor pair whose sum is –11 (the coefficient of the x term).

$$1 + 18 = 19$$
$$-1 + (-18) = -19$$
$$2 + 9 = 11$$
$$-2 + (-9) = -11$$
$$3 + 6 = 9$$
$$-3 + (-6) = -9$$

So, factor using the numbers –2 and –9, as follows:

$$x^2 - 11x + 18 = (x - 2)(x - 9)$$

908. $\quad (x - 4)(x + 5)$

Begin by generating a list of all possible factor pairs of integers (both negative and positive) that multiply to –20 (the constant).

$$(1)(-20) = -20$$
$$(-1)(20) = -20$$
$$(2)(-10) = -20$$
$$(-2)(10) = -20$$
$$(4)(-5) = -20$$
$$(-4)(5) = -20$$

Identify the factor pair whose sum is 1 (the coefficient of the x term).

$$1 + (-20) = -19$$
$$-1 + 20 = 19$$
$$2 + (-10) = -8$$
$$-2 + 10 = 8$$
$$4 + (-5) = -1$$
$$-4 + 5 = 1$$

So, factor using the numbers –4 and 5, as follows:

$$x^2 + x - 20 = (x - 4)(x + 5)$$

909. $(x+2)(x-12)$

Begin by generating a list of all possible factor pairs of integers (both negative and positive) that multiply to –24 (the constant).

$$(1)(-24)=-24$$
$$(-1)(24)=-24$$
$$(2)(-12)=-24$$
$$(-2)(12)=-24$$
$$(3)(-8)=-24$$
$$(-3)(8)=-24$$
$$(4)(-6)=-24$$
$$(-4)(6)=-24$$

Identify the factor pair whose sum is –10 (the coefficient of the x term).

$$1+(-24)=-23$$
$$-1+24=23$$
$$2+(-12)=-10$$
$$-2+12=10$$
$$3+(-8)=-5$$
$$-3+8=5$$
$$4+(-6)=-2$$
$$-4+6=2$$

So, factor using the numbers 2 and –12, as follows:

$$x^2-10x-24=(x+2)(x-12)$$

910. $\dfrac{1}{x+1}$

Begin by factoring x out of the denominator.

$$\frac{x}{x^2+x}=\frac{x}{x(x+1)}$$

Now, cancel a factor of x from both the numerator and denominator.

$$=\frac{1}{x+1}$$

911. x

Begin by factoring x out of the numerator.

$$\frac{x^2-x}{x-1}=\frac{x(x-1)}{x-1}$$

Now, cancel a factor of $x-1$ from both the numerator and denominator.

$$=x$$

912. $\dfrac{x^2}{3}$

Begin by factoring x^2 out of the numerator and 3 out of the denominator.

$$\frac{x^2 + x^4}{3 + 3x^2} = \frac{x^2\left(1 + x^2\right)}{3\left(1 + x^2\right)}$$

Now, cancel a factor of $1 + x^2$ from both the numerator and denominator.

$$= \frac{x^2}{3}$$

913. $\dfrac{4}{5x^2}$

Begin by factoring $4x^2$ out of the numerator and $5x^4$ out of the denominator.

$$\frac{8x^5 + 4x^2}{10x^7 + 5x^4} = \frac{4x^2\left(2x^3 + 1\right)}{5x^4\left(2x^3 + 1\right)}$$

Now, cancel a factor of $2x^3 + 1$ from both the numerator and denominator.

$$= \frac{4x^2}{5x^4}$$

Additionally, you can cancel a factor of x^2 from both the numerator and the denominator.

$$= \frac{4}{5x^2}$$

914. $\dfrac{2x}{5}$

Begin by factoring $2x$ out of the numerator and 5 out of the denominator.

$$\frac{2x^3 + 6x^2 + 8x}{5x^2 + 15x + 20} = \frac{2x\left(x^2 + 3x + 4\right)}{5\left(x^2 + 3x + 4\right)}$$

Now, cancel a factor of $x^2 + 3x + 4$ from both the numerator and denominator.

$$= \frac{2x}{5}$$

915. $x - 2$

Begin by factoring the numerator as the difference of squares.

$$\frac{x^2 - 4}{x + 2} = \frac{(x + 2)(x - 2)}{x + 2}$$

Now, cancel a factor of $x + 2$ from both the numerator and denominator.

$$= x - 2$$

916. $\dfrac{x-y}{2}$

Begin by factoring the numerator as the difference of squares.

$$\frac{x^2 - y^2}{2x + 2y} = \frac{(x+y)(x-y)}{2x+2y}$$

Next, factor out the GCF, 2, in the denominator.

$$= \frac{(x+y)(x-y)}{2(x+y)}$$

Now, cancel a factor of $x + y$ from both the numerator and denominator.

$$= \frac{x-y}{2}$$

917. $\dfrac{2x-5}{4}$

Begin by factoring the numerator as the difference of squares.

$$\frac{4x^2 - 25}{8x + 20} = \frac{(2x+5)(2x-5)}{8x+20}$$

Now, factor out the GCF, 4, in the denominator.

$$= \frac{(2x+5)(2x-5)}{4(2x+5)}$$

Finally, cancel a factor of $2x + 5$ from both the numerator and denominator.

$$= \frac{2x-5}{4}$$

918. $\dfrac{1}{16x(x+2)}$

To begin, factor out the GCF, $16x$, from the denominator.

$$\frac{x-2}{16x^3 - 64x} = \frac{x-2}{16x(x^2 - 4)}$$

Next, factor $x^2 - 4$ in the denominator as the difference of squares.

$$= \frac{x-2}{16x(x+2)(x-2)}$$

Now, cancel a factor of $x - 2$ from both the numerator and denominator.

$$= \frac{1}{16x(x+2)}$$

919. $\dfrac{x+6}{x-1}$

To begin, factor the numerator as the difference of squares.

$$\frac{x^2 - 36}{x^2 - 7x + 6} = \frac{(x+6)(x-6)}{x^2 - 7x + 6}$$

Next, factor the quadratic expression in the denominator.

$$\frac{(x+6)(x-6)}{(x-1)(x-6)}$$

Now, cancel a factor of $x-6$ from both the numerator and denominator.

$$=\frac{x+6}{x-1}$$

920. $\dfrac{x+10}{x-3}$

To begin, factor the quadratic expression in the numerator.

$$\frac{x^2+12x+20}{x^2-x-6}=\frac{(x+2)(x+10)}{x^2-x-6}$$

Next, factor the quadratic expression in the denominator.

$$=\frac{(x+2)(x+10)}{(x+2)(x-3)}$$

Now, cancel a factor of $x+2$ from both the numerator and denominator.

$$=\frac{x+10}{x-3}$$

921. **i. 8, ii. 12, iii. 9, iv. 14, v. 11**

i. $6 + 8 = 14$

ii. $21 - 12 = 9$

iii. $7(9) = 63$

iv. $14 \div 1 = 14$

v. $99 \div 11 = 9$

922. **i. 49, ii. 112, iii. 45, iv. 76, v. 247**

i. $117 - 68 = 49$

ii. $29 + 83 = 112$

iii. $585 \div 13 = 45$

iv. $3{,}116 \div 41 = 76$

v. $19 \times 13 = 247$

923. 12

Begin by testing $x = 10$.

$$9x+14$$
$$=9(10)+14$$
$$=90+14=104$$

Because $104 < 122$, you know that $x = 10$ is a little low, so try $x = 11$.

$$9(11) + 14$$
$$= 99 + 14 = 113$$

This is still a little low, so try $x = 12$.

$$9(12) + 14$$
$$= 108 + 14 = 122$$

924. 29

Begin by testing $x = 25$.

$$30x + 115$$
$$= 30(25) + 115$$
$$= 750 + 115 = 865$$

This is low, so try $x = 30$.

$$30(30) + 115$$
$$= 900 + 115 = 1,015$$

This is just a little high, so try $x = 29$.

$$30(29) + 115$$
$$= 870 + 115 = 985$$

Therefore, $x = 29$.

925. 5

Begin by adding 3 to each side of the equation.

$$6x - 3 = 27$$
$$ +3 \quad +3$$
$$6x = 30$$

Now, divide both sides by 6.

$$\frac{6x}{6} = \frac{30}{6}$$
$$x = 5$$

Therefore, $x = 5$.

926. 7

Begin by subtracting $9n$ from each side of the equation.

$$9n + 14 = 11n$$
$$-9n \quad\quad -9n$$
$$14 = 2n$$

Now, divide both sides by 2.

$$\frac{14}{2} = \frac{2n}{2}$$

$$7 = n$$

Therefore, $n = 7$.

927.   −3

Begin by subtracting v from both sides of the equation.

$$v + 18 = -5v$$

$$\underline{-v \quad\quad -v}$$

$$18 = -6v$$

Now, divide both sides by −6.

$$\frac{18}{-6} = \frac{-6v}{-6}$$

$$-3 = v$$

Therefore, $v = -3$.

928. $\frac{1}{3}$

Begin by subtracting $3k$ from both sides of the equation.

$$9k = 3k + 2$$

$$\underline{-3k \quad -3k}$$

$$6k = 2$$

Now, divide both sides by 6.

$$\frac{6k}{6} = \frac{2}{6}$$

$$k = \frac{1}{3}$$

929.   9

Begin by subtracting $2y$ from both sides of the equation.

$$2y + 7 = 3y - 2$$

$$\underline{-2y \quad\quad -2y}$$

$$7 = y - 2$$

Now, add 2 to both sides.

$$7 = y - 2$$

$$\underline{+2 \quad +2}$$

$$9 = y$$

Therefore, $y = 9$.

930. 16

Begin by subtracting m from both sides of the equation.

$$m + 24 = 3m - 8$$
$$\underline{-m \qquad -m}$$
$$24 = 2m - 8$$

Now, add 8 to both sides.

$$24 = 2m - 8$$
$$\underline{+8 \qquad +8}$$
$$32 = 2m$$

Finally, divide both sides by 2.

$$\frac{32}{2} = \frac{2m}{2}$$
$$16 = m$$

Therefore, $m = 16$.

931. $-3\frac{1}{3}$

Begin by subtracting $-7a$ from both sides of the equation.

$$7a + 7 = 13a + 27$$
$$7 = 6a + 27$$

Now, subtract 27 from both sides; then divide by 6.

$$-20 = 6a$$
$$\frac{-20}{6} = \frac{6a}{6}$$
$$-\frac{10}{3} = a$$
$$-3\frac{1}{3} = a$$

932. −5

Begin by simplifying the equation by combining like terms.

$$3h - 2h + 15 = 5h - 7h$$
$$h + 15 = -2h$$

Now, isolate h and solve.

$$15 = -3h$$
$$\frac{15}{-3} = \frac{-3h}{-3}$$
$$-5 = h$$

---

933. **11**

Begin by simplifying the equation by combining like terms.
$$6x + 4 + 2x = 3 + 9x - 10$$
$$8x + 4 = 9x - 7$$

Now, isolate x and solve.
$$4 = x - 7$$
$$11 = x$$

---

934. **5**

Begin by subtracting $2.3w$ from both sides of the equation.
$$2.3w + 7 = 3.7w$$
$$7 = 1.4w$$

Now, divide both sides by 1.4.
$$\frac{7}{1.4} = \frac{1.4w}{1.4}$$
$$5 = w$$

---

935. **3.5**

Begin by adding $1.9p$ to both sides of the equation.
$$-1.9p + 7 = 2.1p - 7$$
$$7 = 4p - 7$$

Now, add 7 to both sides and then divide by 4.
$$14 = 4p$$
$$\frac{14}{4} = \frac{4p}{4}$$
$$3.5 = p$$

---

936. **$0.0\overline{27}$**

Begin by simplifying the equation by combining like terms.
$$0.8j - 2.4j + 1 = 9.4j + 0.7$$
$$-1.6j + 1 = 9.4j + 0.7$$

Now, add $1.6j$ to both sides and then subtract 0.7 from both sides.
$$1 = 11j + 0.7$$
$$0.3 = 11j$$

Finally, divide both sides by 11.
$$0.0\overline{27} = j$$

937. $-\dfrac{9}{4}$

To begin, simplify each side of the equation, removing parentheses by distributing.

$$3 + (x - 1) = 2x - (5x + 7)$$
$$3 + x - 1 = 2x - 5x - 7$$

Next, combine like terms on each side of the equation; then isolate and solve for x.

$$2 + x = -3x - 7$$
$$2 + 4x = -7$$
$$4x = -9$$
$$x = -\dfrac{9}{4}$$

938. -6

To begin, simplify each side of the equation, removing parentheses by distributing.

$$7u - (10 - 3u) = 5(3u + 4)$$
$$7u - 10 + 3u = 15u + 20$$

Next, combine like terms on each side of the equation; then isolate and solve for u.

$$10u - 10 = 15u + 20$$
$$-10 = 5u + 20$$
$$-30 = 5u$$
$$-6 = u$$

939. $-\dfrac{2}{21}$

To begin, simplify each side of the equation, removing parentheses by distributing.

$$-(2k - 6) = 5(1 + 8k) + 5$$
$$-2k + 6 = 5 + 40k + 5$$

Next, combine like terms on each side of the equation; then isolate and solve for k.

$$-2k + 6 = 10 + 40k$$
$$6 = 10 + 42k$$
$$-4 = 42k$$
$$-\dfrac{4}{42} = k$$
$$-\dfrac{2}{21} = k$$

940. $\dfrac{1}{2}$

To begin, simplify each side of the equation, removing parentheses by distributing.

$$6x(3 + 3x) + 39 = 9x(11 + 2x) - 3x$$
$$18x + 18x^2 + 39 = 99x + 18x^2 - 3x$$

Next, subtract $18x^2$ from each side of the equation; then combine like terms.

$$18x + 39 = 99x - 3x$$

$$18x + 39 = 96x$$

Isolate x.

$$39 = 78x$$

$$\frac{39}{78} = x$$

$$\frac{1}{2} = x$$

941.   2

To begin, distribute on the left side of the equation.

$$1.3(5v) = v + 11$$

$$6.5v = v + 11$$

Simplify and solve for v.

$$6.5v = v + 11$$

$$5.5v = 11$$

$$v = 2$$

942.   1.9

To begin, simplify each side of the equation, removing parentheses by distributing.

$$0.2(15y + 2) = 0.5(8y - 3)$$

$$3y + 0.4 = 4y - 1.5$$

Isolate y.

$$0.4 = y - 1.5$$

$$1.9 = y$$

943.   −15.75

To begin, simplify the left side of the equation, removing the parentheses by distributing.

$$1.75(44m + 36) = 73m$$

$$77m + 63 = 73m$$

Simplify and solve for m.

$$63 = -4m$$

$$\frac{63}{-4} = m$$

$$-15.75 = m$$

944. −3.2

To begin, simplify each side of the equation, removing parentheses by distributing.

$$1.8(3n-5) = 3(4.3n+5)$$
$$5.4n - 9 = 12.9n + 15$$

Isolate n.

$$-9 = 7.5n + 15$$
$$-24 = 7.5n$$
$$-3.2 = n$$

945. −2

To begin, simplify each side of the equation, removing parentheses by distributing.

$$4.4(3s+7) = 4s + 4(s-0.2) - 13.5s - 5.8$$
$$13.2s + 30.8 = 4s + 4s - 0.8 - 13.5s - 5.8$$

Simplify and solve for s.

$$13.2s + 30.8 = -5.5s - 6.6$$
$$18.7s + 30.8 = -6.6$$
$$18.7s = -37.4$$
$$s = -2$$

946. 42

Multiply both sides of the equation by 6.

$$\frac{1}{6}n = 7$$
$$6\left(\frac{1}{6}n\right) = 6(7)$$
$$n = 42$$

947. 44

Multiply both sides of the equation by $\frac{11}{2}$.

$$\frac{2}{11}w = 8$$
$$\frac{11}{2}\left(\frac{2}{11}w\right) = \frac{11}{2}(8)$$
$$w = \frac{88}{2}$$
$$w = 44$$

948. $\frac{20}{27}$

Multiply both sides of the equation by $\frac{4}{3}$.

$$\frac{3}{4}y = \frac{5}{9}$$

$$\frac{4}{3}\left(\frac{3}{4}y\right) = \frac{4}{3}\left(\frac{5}{9}\right)$$

$$y = \frac{20}{27}$$

949. $-\frac{7}{2}$

To begin, cross-multiply to remove the fractions.

$$\frac{q}{7} = \frac{q-1}{9}$$

$$9q = 7(q-1)$$

Simplify and solve for q.

$$9q = 7q - 7$$

$$2q = -7$$

$$q = -\frac{7}{2}$$

950. $-\frac{1}{3}$

To begin, cross-multiply to remove the fractions.

$$\frac{c+2}{5} = \frac{1-3c}{6}$$

$$6(c+2) = 5(1-3c)$$

Simplify and solve for c.

$$6c + 12 = 5 - 15c$$

$$21c + 12 = 5$$

$$21c = -7$$

$$c = -\frac{1}{3}$$

951. $-\frac{5}{2}$

To begin, cross-multiply to remove the fractions.

$$\frac{t-10}{3t} = \frac{t}{3t+6}$$

$$(t-10)(3t+6) = 3t^2$$

"FOIL" the left side of the equation; then subtract $3t^2$ from both sides.

$$3t^2 + 6t - 30t - 60 = 3t^2$$

$$6t - 30t - 60 = 0$$

Combine like terms and isolate t.

$$-24t - 60 = 0$$
$$-24t = 60$$
$$t = \frac{60}{-24}$$
$$t = -\frac{5}{2}$$

952. $\quad -\frac{1}{7}$

To begin, cross-multiply to remove the fractions.

$$\frac{z+1}{z-2} = \frac{z-1}{z+3}$$
$$(z+1)(z+3) = (z-1)(z-2)$$

"FOIL" both sides of the equation; then subtract z^2 from both sides.

$$z^2 + 3z + z + 3 = z^2 - z - 2z + 2$$
$$3z + z + 3 = -z - 2z + 2$$

Simplify and isolate z.

$$4z + 3 = -3z + 2$$
$$7z + 3 = 2$$
$$7z = -1$$
$$z = -\frac{1}{7}$$

953. $\quad \frac{4}{17}$

To begin, cross-multiply to remove the fractions.

$$\frac{3b-4}{6b} = \frac{2b-5}{4b+1}$$
$$(3b-4)(4b+1) = 6b(2b-5)$$

"FOIL" the left side of the equation and distribute the right side; then subtract $12b^2$ from both sides.

$$(3b-4)(4b+1) = 6b(2b-5)$$
$$12b^2 + 3b - 16b - 4 = 12b^2 - 30b$$
$$3b - 16b - 4 = -30b$$

Simplify and isolate b.

$$-13b - 4 = -30b$$
$$-4 = -17b$$
$$\frac{-4}{-17} = \frac{-17b}{-17}$$
$$\frac{4}{17} = b$$

954. 18

To begin, add the two terms on the left side of the equation.

$$p + \frac{p}{9} = 20$$

$$\frac{9p}{9} + \frac{p}{9} = 20$$

$$\frac{10p}{9} = 20$$

Multiply both sides by 9 and solve for p.

$$10p = 180$$

$$p = 18$$

955. 6

To begin, use cross-multiplication techniques to add the fractions on the left side of the equation.

$$\frac{d}{2} + \frac{d}{3} = 5$$

$$\frac{3d + 2d}{6} = 5$$

$$\frac{5d}{6} = 5$$

Multiply both sides of the equation by 6 and then isolate d.

$$6\left(\frac{5d}{6}\right) = 6(5)$$

$$5d = 30$$

$$d = 6$$

956. $\frac{8}{5}$

To begin, use cross-multiplication techniques to add the fractions on the left side of the equation.

$$\frac{3s}{2} + \frac{3s}{4} = s + 2$$

$$\frac{12s + 6s}{8} = s + 2$$

Multiply both sides of the equation by 8 and then simplify and isolate s.

$$12s + 6s = 8(s + 2)$$

$$18s = 8s + 16$$

$$10s = 16$$

$$s = \frac{16}{10}$$

$$s = \frac{8}{5}$$

957. $\dfrac{35}{13}$

To begin, use cross-multiplication techniques to add the fractions on the left side of the equation.

$$\frac{2r}{5} + \frac{r+1}{6} = r - 1$$

$$\frac{12r + 5(r+1)}{30} = r - 1$$

Multiply both sides of the equation by 30 and then simplify.

$$\frac{12r + 5(r+1)}{30} = r - 1$$

$$12r + 5(r+1) = 30(r-1)$$

$$12r + 5r + 5 = 30r - 30$$

$$17r + 5 = 30r - 30$$

Isolate r.

$$5 = 13r - 30$$

$$35 = 13r$$

$$\frac{35}{13} = r$$

958. 3

To begin, increase the terms of the first fractions by 2 (so that the common denominator is 4); then multiply both sides of the equation by 4 to eliminate the fractions.

$$\frac{j}{2} + \frac{3}{4} = \frac{3j}{4}$$

$$\frac{2j}{4} + \frac{3}{4} = \frac{3j}{4}$$

$$2j + 3 = 3j$$

Isolate j to solve.

$$3 = j$$

959. $-\dfrac{4}{3}$

To begin, change all three fractions so that they have denominators of 8; then multiply both sides of the equation by 8 to eliminate the fractions.

$$\frac{1}{2} + \frac{5k}{8} = \frac{k}{4}$$

$$\frac{4}{8} + \frac{5k}{8} = \frac{2k}{8}$$

$$4 + 5k = 2k$$

Isolate k to solve.

$$4 = -3k$$

$$-\frac{4}{3} = k$$

960. $-\dfrac{1}{10}$

To begin, increase the terms of all three fractions to change all denominators to 12. Then multiply both sides of the equation by 12 to eliminate the fractions.

$$\frac{8a}{3} + \frac{1}{4} = \frac{a}{6}$$

$$\frac{32a}{12} + \frac{3}{12} = \frac{2a}{12}$$

$$32a + 3 = 2a$$

Isolate a to solve.

$$3 = -30a$$

$$\frac{3}{-30} = \frac{-30a}{-30}$$

$$-\frac{1}{10} = a$$

961. 6

To begin, increase the terms of all three fractions to change all denominators to 60; then multiply both sides of the equation by 60 to eliminate the fractions.

$$\frac{h}{30} + \frac{3h}{20} = \frac{11}{10}$$

$$\frac{2h}{60} + \frac{9h}{60} = \frac{66}{60}$$

$$2h + 9h = 66$$

Isolate h to solve.

$$11h = 66$$

$$h = 6$$

962. -11

Begin by changing all terms to a denominator of 9; then multiply both sides of the equation by 9 to eliminate the fractions.

$$\frac{2}{3}k + 5 = \frac{1}{9}(2k + 1)$$

$$\frac{6k}{9} + \frac{45}{9} = \frac{2k + 1}{9}$$

$$6k + 45 = 2k + 1$$

Isolate k.

$$4k + 45 = 1$$

$$4k = -44$$

$$k = -11$$

963. $\frac{4}{3}$

Begin by changing all terms to a denominator of 8; then multiply both sides of the equation by 8 to eliminate the fractions.

$$\frac{1}{2}(3-y)+\frac{1}{4}=\frac{1}{8}(2y+6)$$

$$\frac{4}{8}(3-y)+\frac{2}{8}=\frac{1}{8}(2y+6)$$

$$4(3-y)+2=1(2y+6)$$

Distribute to remove parentheses:

$$12-4y+2=2y+6$$

Isolate y:

$$14-4y=2y+6$$

$$8-4y=2y$$

$$8=6y$$

$$\frac{8}{6}=y$$

$$\frac{4}{3}=y$$

964. 7

Divide both sides of the equation by x^3.

$$5x^4=35x^3$$

$$\frac{5x^4}{x^3}=\frac{35x^3}{x^3}$$

$$5x=35$$

Now, divide both sides by 5.

$$x=7$$

965. $\pm\frac{\sqrt{5}}{5}$

Divide both sides of the equation by x^4.

$$45x^6=9x^4$$

$$\frac{45x^6}{x^4}=\frac{9x^4}{x^4}$$

$$45x^2=9$$

Next, divide both sides by 45.

$$\frac{45x^2}{45}=\frac{9}{45}$$

$$x^2=\frac{1}{5}$$

Now, take the square root of both sides.

$$\sqrt{x^2} = \pm\sqrt{\frac{1}{5}}$$

$$x = \pm\frac{\sqrt{1}}{\sqrt{5}}$$

$$x = \pm\frac{1}{\sqrt{5}} = \pm\frac{\sqrt{5}}{5}$$

966. **5 and –5**

Isolate x and solve.

$$x^2 - 25 = 0$$

$$x^2 = 25$$

$$x = 5, -5$$

967. **7 and –9**

Begin by factoring the left side of the equation.

$$x^2 + 2x - 63 = 0$$

$$(x + 9)(x - 7) = 0$$

Now, split this equation into two separate equations and solve them.

$$x + 9 = 0 \qquad x - 7 = 0$$

$$x = -9 \qquad x = 7$$

968. **1 and –8**

Begin by moving all terms to one side of the equation.

$$8x^2 + 7x = 7x^2 + 8$$

$$x^2 + 7x = 8$$

$$x^2 + 7x - 8 = 0$$

Now, factor the left side of the equation.

$$(x + 8)(x - 1) = 0$$

Now, split this equation into two separate equations and solve them.

$$x + 8 = 0 \qquad x - 1 = 0$$

$$x = -8 \qquad x = 1$$

969. **6 and 7**

Begin by distributing on both sides of the equation to remove the parentheses; then move all terms to one side.

$$7\left(x^2 - x + 3\right) = 3\left(2x^2 + 2x - 7\right)$$
$$7x^2 - 7x + 21 = 6x^2 + 6x - 21$$
$$x^2 - 13x + 42 = 0$$

Factor the left side of the equation.

$$(x - 6)(x - 7) = 0$$

Now, split this equation into two separate equations and solve them.

$$x - 6 = 0 \qquad x - 7 = 0$$
$$x = 6 \qquad x = 7$$

970. **–3 and –5**

Begin by cross-multiplying to remove the fractions.

$$\frac{2x + 1}{15} = \frac{x^2 - 1}{8x}$$
$$8x(2x + 1) = 15\left(x^2 - 1\right)$$

Now, distribute on both sides of the equation to remove the parentheses; then move all terms to one side.

$$16x^2 + 8x = 15x^2 - 15$$
$$16x^2 + 8x + 15 = 15x^2$$
$$x^2 + 8x + 15 = 0$$

Now, factor the left side of the equation.

$$(x + 3)(x + 5) = 0$$

Now, split this equation into two separate equations and solve them.

$$x + 3 = 0 \qquad x + 5 = 0$$
$$x = -3 \qquad x = -5$$

971. **2d + 1,000**

The amount d doubles to $2d$, and then increases by 1,000 to $2d + 1,000$.

972. **3c – 60**

The day begins with c chairs. Then 20 chairs are removed, bringing the number to $c - 20$. After that, this number is tripled, which brings the number to

$$3(c - 20) = 3c - 60$$

973. $p - 234$

Penny starts with p pennies. She then removes 300 pennies, bringing the total to $p - 300$ pennies. The next day, she adds back in 66 pennies, so the total becomes

$$p - 300 + 66$$

You can simplify this amount as follows:

$$= p - 234$$

974. $t - 2$

The temperature begins at t degrees and then changes as follows:

$$t + 5 + 2 - 3 - 6 = t - 2$$

975. $6w + 12$

The puppy's weight begins at w. It triples to $3w$, then increases by 6 pounds to $3w + 6$, and finally doubles to

$$2(3w + 6) = 6w + 12$$

976. $2k + 57$

Kyle has k baseball cards. Randy has half as many, so Randy has $\frac{k}{2}$ cards. And Jacob has 57 more cards than Randy, so Jacob has $\frac{k}{2} + 57$ cards. Add these up as follows:

$$k + \left(\frac{k}{2}\right) + \left(\frac{k}{2} + 57\right)$$

You can further simplify this by combining the three k terms.

$$= 2k + 57$$

977. $0.72s + 425$

The school currently has s students. The number of graduating students is $0.28s$. When these students leave, the number of remaining students will be

$$s - 0.28s = 0.72s$$

Additionally, 425 new students will be at the school, so this number will increase to $0.72s + 425$.

978. $5m + 10$

Millie walked m miles the first day, $m + 1$ miles the second day, $m + 2$ miles the third day, $m + 3$ miles the fourth day, and $m + 4$ miles the fifth day. The sum of these numbers is

$$m + m + 1 + m + 2 + m + 3 + m + 4$$

Combine like terms to simplify.

$$= 5m + 10$$

979. $4n + 12$

Every consecutive odd number is exactly two greater than the preceding one. So, you can represent the four numbers as n, $n + 2$, $n + 4$, and $n + 6$. Thus, the sum of these numbers is:

$$n + n + 2 + n + 4 + n + 6$$

Simplify as follows:

$$= 4n + 12$$

980. 4

Let x equal the number. Then, set up and solve the following equation:

$$6x - 1 = 23$$
$$6x = 24$$
$$x = 4$$

981. 4

Let x equal the number. Then, set up and solve the following equation:

$$3x = x + 8$$
$$2x = 8$$
$$x = 4$$

982. 8

Let x equal the number. Then, set up and solve the following equation:

$$5x = 3x + 16$$
$$2x = 16$$
$$x = 8$$

983. −2

Let x equal the number. Then, set up and solve the following equation:

$$2x + 7 = 3x + 9$$
$$7 = x + 9$$
$$-2 = x$$

984. 7

Let x equal the number. Then, set up and solve the following equation:

$$2(x+6)=5x-9$$
$$2x+12=5x-9$$
$$12=3x-9$$
$$21=3x$$
$$7=x$$

985. 17

Let x equal the number. Then, set up and solve the following equation:

$$2(x+3)=4(x-7)$$
$$2x+6=4x-28$$
$$6=2x-28$$
$$34=2x$$
$$17=x$$

986. 23

Let x equal the number. Then, set up and solve the following equation:

$$\frac{x+1}{3}=\frac{x-7}{2}$$
$$3(x-7)=2(x+1)$$
$$3x-21=2x+2$$
$$x-21=2$$
$$x=23$$

987. 3

Let x equal the number. Then, set up and solve the following equation:

$$\frac{2x-1}{5}=\frac{x}{3}$$
$$3(2x-1)=5x$$
$$6x-3=5x$$
$$x-3=0$$
$$x=3$$

988. 7.25

Let x equal the number. Then, set up and solve the following equation:

$$4(x-6.5)) = x - 4.25$$
$$4x - 26 = x - 4.25$$
$$3x - 26 = -4.25$$
$$3x = 21.75$$
$$x = 7.25$$

989. −11

Let x equal the number. Then, set up and solve the following equation:

$$\frac{x+13}{4} = x + 11.5$$
$$x + 13 = 4(x + 11.5)$$
$$x + 13 = 4x + 46$$
$$13 = 3x + 46$$
$$-33 = 3x$$
$$-11 = x$$

990. 3.6

Let x equal the number. Then, set up and solve the following equation:

$$\frac{x+1.5}{3} = 2x - 5.5$$
$$x + 1.5 = 3(2x - 5.5)$$
$$x + 1.5 = 6x - 16.5$$
$$1.5 = 5x - 16.5$$
$$18 = 5x$$
$$3.6 = x$$

991. 4

Let x equal the number. Then, set up and solve the following equation:

$$\frac{x^2 - 12}{4} = \left(\frac{x}{2} - 1\right)^2$$

$$\frac{x^2 - 12}{4} = \left(\frac{x-2}{2}\right)^2$$

$$\frac{x^2 - 12}{4} = \frac{(x-2)(x-2)}{4}$$

$$x^2 - 12 = (x-2)(x-2)$$

$$x^2 - 12 = x^2 - 4x + 4$$

$$-12 = -4x + 4$$

$$-16 = -4x$$

$$4 = x$$

992. 16

Let x equal the number. Then, set up and solve the following equation:

$$\frac{3}{4}x - 2 = \frac{2}{5}(x + 9)$$

$$\frac{3x}{4} - 2 = \frac{2x + 18}{5}$$

$$\frac{3x}{4} - \frac{8}{4} = \frac{2x + 18}{5}$$

$$\frac{3x - 8}{4} = \frac{2x + 18}{5}$$

$$5(3x - 8) = 4(2x + 18)$$

$$15x - 40 = 8x + 72$$

$$7x - 40 = 72$$

$$7x = 112$$

$$x = 16$$

993. $11

Let p = the number of dollars that Peter has. Then Lucy has $p + 5$ dollars. Together, they have $27, so

$$p + p + 5 = 27$$

Solve for p.

$$2p + 5 = 27$$

$$2p = 22$$

$$p = 11$$

994. **$170**

Let m = the cost of the MP3 player in dollars. Then $2m$ is the cost of the cellphone and $4m$ is the cost of the laptop computer. So, you can set up the following equation:

$$m + 2m + 4m = 1,190$$

Simplify and solve for m.

$$7m = 1,190$$
$$m = 170$$

Therefore, the MP3 player cost $170.

995. **5 years old**

Let j be Jane's age. Then, Cody's age is $j + 8$ and Brent's age is $2j$. Cody is 3 years older than Brent, so you can set up the following equation:

$$\text{Brent} + 3 = \text{Cody}$$
$$2j + 3 = j + 8$$

Simplify and solve for j.

$$2j = j + 5$$
$$j = 5$$

Therefore, Jane is 5 years old.

996. **2 hours and 20 minutes**

Let x be the number of minutes that the class takes. So, the teacher spends $\frac{x}{2}$ minutes going over homework problems and $\frac{x}{5}$ minutes reviewing for a test. Thus, you can set up the following equation:

$$\frac{x}{2} + \frac{x}{5} + 42 = x$$

Raise the terms of every term in this equation to a denominator of 10, then drop the denominators.

$$\frac{5x}{10} + \frac{2x}{10} + \frac{420}{10} = \frac{10x}{10}$$
$$5x + 2x + 420 = 10x$$

Simplify and solve for x.

$$7x + 420 = 10x$$
$$420 = 3x$$
$$140 = x$$

Therefore, the class is 140 minutes long, which equals 2 hours and 20 minutes.

997.   35

Let x be the first number. Then the other four numbers are $x + 1$, $x + 2$, $x + 3$, and $x + 4$. Thus, you can set up the following equation:

$$x + x + 1 + x + 2 + x + 3 + x + 4 = 165$$

Simplify and solve for x.

$$5x + 10 = 165$$
$$5x = 155$$
$$x = 31$$

Thus, the five numbers are 31, 32, 33, 34, and 35. So the greatest is 35.

998.   78

Let y be the number of yellow marbles in the jar. Then, the jar contains $3y$ orange marbles, $y + 6$ blue marbles, and $2(y + 6)$ red marbles. So you can set up the following equation:

$$y + 3y + (y + 6) + 2(y + 6) = 172$$

Simplify and solve for y.

$$y + 3y + y + 6 + 2y + 12 = 172$$
$$7y + 18 = 172$$
$$7y = 154$$
$$y = 22$$

Therefore, the jar contains 22 yellow marbles, so it contains 28 blue marbles and 56 red marbles. Therefore, it contains 22 + 56 = 78 yellow and red marbles.

999.   50 mph

Let s be the speed of the southbound train. Then, $2s$ is the speed of the northbound train and $2s - 10$ is the speed of the eastbound train.

$$s + 40 = 2s - 10$$
$$40 = s - 10$$
$$50 = s$$

Therefore, the southbound train is traveling at 50 mph.

1000. **$300**

Let k equal the number of dollars that Ken has. Then Walter has $k - 100$ dollars. So, you can set up the following equation:

$$3k + \frac{k-100}{2} = 2(k + k - 100)$$

Simplify on the right and increase the terms of each term to a denominator of 2; then drop the denominators.

$$3k + \frac{k-100}{2} = 4k - 200$$

$$\frac{6k}{2} + \frac{k-100}{2} = \frac{8k-400}{2}$$

$$6k + k - 100 = 8k - 400$$

Simplify and solve for k.

$$7k - 100 = 8k - 400$$

$$-100 = k - 400$$

$$300 = k$$

Therefore, Ken has $300.

1001. **12**

Let d be Damar's age now. So Jessica's age now is $2d$. Three years ago, Damar's age was $d - 3$ and Jessica's age was $2d - 3$. And at that time, Jessica was 3 times as old as Damar, so

$$2d - 3 = 3(d - 3)$$

Solve for d.

$$2d - 3 = 3d - 9$$

$$-3 = d - 9$$

$$6 = d$$

Thus, Damar is 6 years old right now. Jessica is twice his age, so she is now 12 years old.

Index

Workspace

Workspace

Workspace

Workspace

Workspace

Workspace

Workspace

Workspace

Workspace

Workspace

Workspace

Workspace

Workspace

Workspace

Workspace

Workspace

Workspace

Workspace

Workspace

e & Mac

For Dummies,
Edition
1-118-49823-1

ne 5 For Dummies,
Edition
1-118-35201-4

Book For Dummies,
Edition
1-118-20920-2

Mountain Lion
Dummies
1-118-39418-2

ging & Social Media

book For Dummies,
dition
1-118-09562-1

Blogging
Dummies
1-118-03843-7

rest For Dummies
1-118-32800-2

Press For Dummies,
dition
1-118-38318-6

ess

modities For Dummies,
Edition
1-118-01687-9

ting For Dummies,
dition
0-470-90545-6

**Personal Finance
For Dummies, 7th Edition**
978-1-118-11785-9

**QuickBooks 2013
For Dummies**
978-1-118-35641-8

**Small Business Marketing
Kit For Dummies,
3rd Edition**
978-1-118-31183-7

Careers

Job Interviews
For Dummies, 4th Edition
978-1-118-11290-8

Job Searching with
Social Media
For Dummies
978-0-470-93072-4

Personal Branding
For Dummies
978-1-118-11792-7

Resumes For Dummies,
6th Edition
978-0-470-87361-8

Success as a Mediator
For Dummies
978-1-118-07862-4

Diet & Nutrition

Belly Fat Diet For Dummies
978-1-118-34585-6

Eating Clean For Dummies
978-1-118-00013-7

Nutrition For Dummies,
5th Edition
978-0-470-93231-5

Digital Photography

Digital Photography
For Dummies,
7th Edition
978-1-118-09203-3

Digital SLR Cameras &
Photography For Dummies,
4th Edition
978-1-118-14489-3

Photoshop Elements 11
For Dummies
978-1-118-40821-6

Gardening

Herb Gardening
For Dummies, 2nd Edition
978-0-470-61778-6

Vegetable Gardening
For Dummies, 2nd Edition
978-0-470-49870-5

Health

Anti-Inflammation Diet
For Dummies
978-1-118-02381-5

Diabetes For Dummies,
3rd Edition
978-0-470-27086-8

Living Paleo For Dummies
978-1-118-29405-5

Hobbies

Beekeeping
For Dummies
978-0-470-43065-1

eBay For Dummies,
7th Edition
978-1-118-09806-6

Raising Chickens
For Dummies
978-0-470-46544-8

Wine For Dummies,
5th Edition
978-1-118-28872-6

Writing Young Adult Fiction
For Dummies
978-0-470-94954-2

Language &
Foreign Language

500 Spanish Verbs
For Dummies
978-1-118-02382-2

English Grammar
For Dummies, 2nd Edition
978-0-470-54664-2

French All-in One
For Dummies
978-1-118-22815-9

German Essentials
For Dummies
978-1-118-18422-6

Italian For Dummies,
2nd Edition
978-1-118-00465-4

e Available in print and e-book formats.

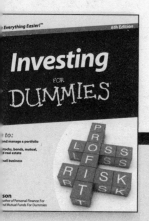

Math & Science

Algebra I For Dummies,
2nd Edition
978-0-470-55964-2

Anatomy and Physiology
For Dummies,
2nd Edition
978-0-470-92326-9

Astronomy For Dummies,
3rd Edition
978-1-118-37697-3

Biology For Dummies,
2nd Edition
978-0-470-59875-7

Chemistry For Dummies,
2nd Edition
978-1-1180-0730-3

Pre-Algebra Essentials
For Dummies
978-0-470-61838-7

Microsoft Office

Excel 2013 For Dummies
978-1-118-51012-4

Office 2013 All-in-One
For Dummies
978-1-118-51636-2

PowerPoint 2013
For Dummies
978-1-118-50253-2

Word 2013 For Dummies
978-1-118-49123-2

Music

Blues Harmonica
For Dummies
978-1-118-25269-7

Guitar For Dummies,
3rd Edition
978-1-118-11554-1

iPod & iTunes
For Dummies,
10th Edition
978-1-118-50864-0

Programming

Android Application
Development For Dummies,
2nd Edition
978-1-118-38710-8

iOS 6 Application
Development For Dummies
978-1-118-50880-0

Java For Dummies,
5th Edition
978-0-470-37173-2

Religion & Inspiration

The Bible For Dummies
978-0-7645-5296-0

Buddhism For Dummies,
2nd Edition
978-1-118-02379-2

Catholicism For Dummies,
2nd Edition
978-1-118-07778-8

Self-Help & Relationships

Bipolar Disorder
For Dummies,
2nd Edition
978-1-118-33882-7

Meditation For Dummies,
3rd Edition
978-1-118-29144-3

Seniors

Computers For Seniors
For Dummies,
3rd Edition
978-1-118-11553-4

iPad For Seniors
For Dummies,
5th Edition
978-1-118-49708-1

Social Security
For Dummies
978-1-118-20573-0

Smartphones & Tablets

Android Phones
For Dummies
978-1-118-16952-0

Kindle Fire HD
For Dummies
978-1-118-42223-6

NOOK HD For Dummies,
Portable Edition
978-1-118-39498-4

Surface For Dummies
978-1-118-49634-3

Test Prep

ACT For Dummies,
5th Edition
978-1-118-01259-8

ASVAB For Dummies,
3rd Edition
978-0-470-63760-9

GRE For Dummies,
7th Edition
978-0-470-88921-3

Officer Candidate Tests
For Dummies
978-0-470-59876-4

Physician's Assistant E:
For Dummies
978-1-118-11556-5

Series 7 Exam
For Dummies
978-0-470-09932-2

Windows 8

Windows 8 For Dummie
978-1-118-13461-0

Windows 8 For Dummie
Book + DVD Bundle
978-1-118-27167-4

Windows 8 All-in-One
For Dummies
978-1-118-11920-4

e Available in print and e-book formats.

Take Dummies with you everywhere you go!

Whether you're excited about e-books, want more from the web, must have your mobile apps, or swept up in social media, Dummies makes everything easier .

...mies products make life easie...

- Software
- Cookware
- Hobbies

- Consumer Electronics
- Crafts

- Videos
- Music
- Games
- and More!

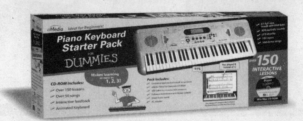

For more information, go to **Dummies.com®** and search the store by catego...

FOR

DUMMIE...

A Wiley B...

For Dummies is a registered trademark of John Wiley & Sons, Inc.